Gravity wave emission from jet systems in the differentially heated rotating annulus experiment

From the Faculty of Mechanical Engineering, Electrical and Energy
Systems
– Chair of Aerodynamics and Fluid Mechanics –
of the Brandenburg Technical University Cottbus-Senftenberg to obtain
the academic degree of Doctor of Engineering

approved dissertation

presented by

Master of Science (M. Sc.)

Costanza RODDA

born on the 6th of March 1988 in Trieste (Italy)

Head of commission:	Prof. Dr. Christoph Egbers (BTU C-S)
Supervisor:	Apl. Prof. Dr. Uwe Harlander (BTU C-S)
Reviewer:	Prof. Dr. Ulrich Achatz (GUF)
Day of the thesis defence:	21st October 2019

Schwerewellenabstrahlung von Strahlströmen im differentiell geheizten rotierenden Annulus Experiment

Von der Fakultät für Maschinenbau, Elektro- und Energiesysteme
-Lehrstuhl für Aerodynamik und Strömungslehre-
der Brandenburgischen Technischen Universität Cottbus-Senftenberg zur
Erlangung des akademischen Grades eines Doktors der
Ingenieurwissenschaften

genehmigte Dissertation

vorgelegt von

Master of Science (M. Sc.)

Costanza Rodda

geboren am 6. März 1988 in Trieste (Italien)

Vorsitzender:	Prof. Dr. Christoph Egbers (BTU C-S)
1. Gutachter:	Apl. Prof. Dr. Uwe Harlander (BTU C-S)
2. Gutachter:	Prof. Dr. Ulrich Achatz (GUF)
Tag der mündlichen Prüfung:	21. Oktober 2019

Bibliografische Information der Deutschen Nationalbibliothek

Die Deutsche Nationalbibliothek verzeichnet diese Publikation in der Deutschen Nationalbibliographie; detaillierte bibliographische Daten sind im Internet über http://dnb.d-nb.de abrufbar.

1. Aufl. - Göttingen: Cuvillier, 2019

  Zugl.: (BTU) Cottbus-Senftenberg, Univ., Diss., 2019

© CUVILLIER VERLAG, Göttingen 2019
  Nonnenstieg 8, 37075 Göttingen
  Telefon: 0551-54724-0
  Telefax: 0551-54724-21
  www.cuvillier.de

  ISBN 978-3-7369-7110-3
  eISBN 978-3-7369-6110-4

Ma il vento riviene,
rincalza, ridonda.
Altra onda s'alza,
nel suo nascimento
più lene.
Palpita, sale,
si gonfia, s'incurva,
s'allunga, propende.
Il dorso ampio splende
come cristallo;
la cima leggera
s'arruffa
come criniera
nivea di cavallo.
Il vento la scavezza.
L'onda si spezza, precipita nel cavo
del solco sonora;
spumeggia, biancheggia,
s'infiora, odora,
travolge la cuora,
trae l'alga e l'ulva;
s'allunga,
rotola, galoppa;
intoppa
in altra cui'l vento
diè tempra diversa;
l'avversa,
la salta, la sormonta,
vi si mesce, s'accresce.
Di spruzzi, di sprazzi,
di fiocchi, d'iridi
ferve nella risacca;
par che di crisopazi
scintilli
e di berilli
vividi a sacca.

Gabriele D'Annunzio–L'onda

Contents

List of Figures

List of Tables

Abstract

Internal gravity waves are fully recognised for the relevant role they play in the atmospheric and oceanic dynamics of our planet. These waves, characterised by small-scale wavelengths, permeate the atmosphere and can lead to energy and momentum flux. When gravity waves propagate upwards, the momentum flux can become divergent due to wave breaking processes and consequently be a source for the large scale flow. Over the past decades, several studies have successfully investigated generation, propagation, and dissipation of gravity waves. Although several phenomena are nowadays reasonably well explained, others still remain unclear and, therefore, an active research topic. One of the least understood aspects is the emission of gravity waves from jet and front systems in the atmosphere, which are regions where significant wave activity has been frequently observed. Although many studies have established the importance of these non-orographic sources, the mechanisms responsible for wave emissions are still not fully understood. The complexity of the three-dimensional flow pattern, where a large number of interacting processes occur, and distribution of the sources over large areas point towards the need for laboratory experiments and idealised numerical simulations. Indeed, experiments and simulations can help with the correct interpretation of the fundamental dynamical processes in a simplified, but yet realistic flow.

In this thesis, we propose an experimental laboratory investigation of gravity waves generated from baroclinic jets and fronts using a differentially heated rotating annulus. This experiment consists of a rotating annular tank made of three concentric compartments filled with water, which is held to a constant warm/cold temperature in the outer/inner ring, respectively. Thanks to the radial temperature difference and the rotation about its axis, resembling each one of the fundamental forcing of the planetary atmospheric circulation, this experiment is a well established model of the atmosphere and it has been used for many years to investigate different aspects of its dynamics. Our study, however, reveals that the classical set-up of this experiment, usually showing an aspect ratio of about one, is not a particularly favourable configuration to investigate atmosphere-like emission of gravity waves from baroclinic jets due to an unrealistic ratio between stratification, measured by the buoyancy frequency N, and rotation, measured by the Coriolis parameter f. This ratio is much greater than one for the atmosphere but less than one for the classical annulus, and this difference is crucial when studying gravity waves. Indeed, no proof of gravity waves emission from jets has been found in the classical experiment thus far. Based on this, we offer two modifications of the original experiment to obtain configurations suitable for our purposes.

The first uses the same geometrical configuration, i.e. a tall and narrow annulus, but with the introduction of salinity stratification, which increases N up to values that reach $2 < N/f < 6$. The second solution follows the numerical simulations by Borchert et al. (2014) where a new shallower tank, called the atmosphere-like annulus, was proposed. The newly built tank has a much larger horizontal diameter which leads to a more atmosphere-like $N/f > 1$.

With the use of the Particle Image Velocimetry measurement technique and temperature sensors, the flow at different depths is investigated. Gravity waves are observed along the baroclinic jet in both experimental set-ups, their properties, together with the conditions for their emission and propagation, are examined in detail. In particular, the regions of the regime diagram for which the gravity waves with the largest amplitude can develop are identified. Subsequently, four possible generation mechanisms—i.e. shear instability, lateral wall instabilities, convection, and spontaneous emission—are analysed and matched with the properties of the waves. Despite the differences between both experimental configurations the gravity wave signature shows many similarities suggesting that similar generation processes occur in both cases. Finally, the energy partition among large- and small-scale wave phenomena is shown. The resemblance of the spectra obtained for the atmosphere-like annulus with the ones measured in the atmosphere is noteworthy, therefore proving that this experimental configuration is well suited for the investigation of multi-scale dynamics.

Zusammenfassung

Interne Schwerewellen spielen eine wichtige Rolle für die atmosphärische und ozeanische Dynamik unseres Planeten. Diese Wellen zeichnen sich durch kurze Wellenlängen aus, durchdringen die Atmosphäre und sorgen für einen Impuls- und Energiefluss. Im Falle einer Ausbreitung nach oben, kann durch Wellenbrechen Impuls und Energie auf die großräumige Strömung übertragen werden. In den letzten Jahrzehnten haben viele Studien die Erzeugung, Ausbreitung und Dissipation von Schwerewellen untersucht und obwohl einige Phänomene heutzutage gut erklärt sind, bleiben andere nach wie vor unklar und daher ein aktives Forschungsthema. Einer der nur wenig verstandenen Aspekte ist die Abstrahlung von internen Schwerewellen in atmosphärischen Strahlströmen und Wetterfronten, in denen häufig signifikante Schwerewellenaktivität beobachtet werden kann. Obwohl viele Studien die Bedeutung dieser nicht orographischen Quellen belegen, sind die Mechanismen, die für die Wellenemission verantwortlich sind, noch nicht vollständig geklärt. Die Komplexität des dreidimensionalen Strömungsfeldes, in dem wichtige Wechselwirkungsprozesse ablaufen und auch die großräumige Verteilung der Quellen für Schwerewellen legen nahe, Laborversuche und idealisierte numerische Simulationen durchzuführen. Diese helfen bei der korrekten Interpretation der grundlegenden dynamischen Prozesse in einem vereinfachten, aber dennoch realistischen Szenario.ö In dieser Arbeit schlage ich eine experimentelle Laboruntersuchung von Schwerewellen vor, die im Bereich von baroklinen Jetströmungen und Fronten in einem differentiell geheizten und rotierenden Annulus erzeugt werden. Dieses Experiment besteht aus einem rotierenden ringförmigen Tank mit drei koaxialen Zylindern. Diese Bereiche sind mit Wasser gefüllt und der Außen- bzw. Innenring kann auf einer konstanten Temperatur (innen kalt und außen warm) gehalten werden. Dank der radialen Temperaturdifferenz und der Drehung des Tanks um seine Symmetrieachse, ist dieses Experiment ein etabliertes Analogon der Atmosphäre und wird seit vielen Jahren verwendet, um verschiedene Aspekte der atmosphärischen Zirkulation zu untersuchen. Unsere Studie zeigt jedoch, dass der klassische Aufbau dieses Experiments, der normalerweise ein Seitenverhältnis von etwa Eins aufweist, keine besonders günstige Konfiguration für die Untersuchung von Schwerewellen darstellt. Dies liegt im Wesentlichen an einem unrealistischen Verhältnisses zwischen der Auftriebsfrequenz N und dem Coriolis-Parameters f. Dieses Verhältnis ist viel grösser als Eins für die Atmosphäre, aber kleiner als Eins für das klassische Annulus Experiment und dieser Unterschied ist entscheidend, wenn Schwerewellen entstehen sollen. Tatsächlich wurde im klassischen Experiment bisher kein Nachweis der Schwerkraftwellabstrahlung gefunden. In der vorliegenden Arbeit biete ich zwei Modifikationen des ursprünglichen Experiments an, um Konfigurationen zu erhalten, die für Schwerewellenabstrahlung geeignet sind. Die erste verwendet dieselbe geometrische Konfiguration, d.h. einen hohen und schmalen Aufbau, jedoch mit einer zusätzlichen Salz-Schichtung, die N/f auf Werte von 2-6 erhöht. Eine zweite, bereits von Borchert et al. (2014) vorgeschlagene Variante besteht aus einem neu gebauten, flachen Tank, genannt atmosphärenähnlicher Annulus, dessen Verhältnis N/f auch ohne zusätzliche Salz-Schichtung größer als Eins ist. Dies kann durch theoretische überlegungen gefunden werden und wurde auch durch numerische Simulationen bestätigt (Borchert et al. 2014). Mit der Particle-Image-Velocimetry Messtechnik und Temperatursensoren untersuche ich in meiner Studie die Strömung in verschiedenen Tiefen. In beiden Versuchsaufbauten werden entlang des baroklinen Jets Schwerewellen beobachtet, deren Eigenschaften sowie die Bedingungen für ihre Emission und Ausbreitung detailliert untersucht werden. Insbesondere werden die Parameterbereiche identifiziert, für die sich eine reproduzierbare Schwerewellenausbreitung finden lässt. Anschließend werden vier mögliche Erzeugungsmechanismen, nämlich Scher- und Grenzschichtinstabilität, Konvektion und spontane Emission analysiert und mit den Eigenschaften der beobachteten Wellen abgeglichen. Trotz der Unterschiede zwischen den beiden genannten Experimentkonfigurationen zeigt deren Schwerewellensignatur viele ähnlichkeiten, die darauf hindeuten, dass in beiden Fällen ein ähnlicher Entstehungsmechanismus vorliegt. Schließlich wird die Energieaufteilung zwischen verschiedenen Wellen und Skalen aus Experimentdaten bestimmt. Die ähnlichkeit der für den atmosphärenähnlichen Annulus gefundenen Energiespektren mit denen der realen Atmosphäre ist bemerkenswert. Letzteres zeigt, dass diese experimentelle Konfiguration für die hier durchgeführte Untersuchung der Multiskalendynamik von Schwerewellen ausgezeichnet geeignet ist.

Chapter 1

Introduction

"Perhaps some day in the dim future it will be possible to advance the computations faster than the weather advances and at a cost less than the saving to mankind due to the information gained. But that is a dream."

— Lewis Fry Richardson (1922) —

1.1 Motivation and Problem Statement

The study of atmospheric general circulation dates back to the 18th century with the pioneering work of Hadley (1735), who first realised that differential heating is the driving mechanism to the circulation. Since then, much effort has been made for explaining the atmospheric motions and developing a theoretical model of the general circulation. A fundamental point for developing such a model is to understand the dominating factors and mechanisms of the general circulation so that the essential physics of the atmospheric circulation is captured. To this purpose, experiments (both simulations and laboratory) analogues of the atmosphere are the perfect way to test and develop simplifying theoretical hypothesis and separate the fundamental mechanisms driving the atmospheric dynamics to other phenomena that can influence and modify the general circulation. Thanks to all the work done in the past decades, nowadays we have a good understanding of the essential physical model of the atmospheric circulation, which can be summarised as a rotating fluid under the forcing of Solar heating and gravity.

The combination of the development of equations describing the motions of the atmosphere and computers powerful enough to numerically solve them, led in the '50s to the first weather forecasts. Nowadays, weather forecasts, climate models and global circulation models are able to reproduce and predict not only the essential atmospheric features, but also other aspects of secondary importance that are, nevertheless, relevant for the dynamics. Indeed, in addition to the driving mechanisms, several other phenomena can come into play and affect the atmospheric dynamics. Among these phenomena, internal gravity waves—which are the focus of this thesis—are known to have a significant impact on the atmospheric circulation, structure, and variability (Fritts and Alexander 2003).

These small-scale waves, with horizontal dimensions in the order of 100 km to 1000 km, own their name to a combination of the location where they occur in the fluid (the interior) and their restoring force (gravity). When these waves have low frequencies and the rotation of the Earth has a stronger influence on them, they are called 'inertia-gravity waves' (IGWs in short). In this thesis, the term internal gravity waves shall be used also for indicating IGWs, regardless of the influence of rotation. Gravity waves develop in any stratified fluid and, therefore, are omnipresent in the oceans and atmosphere. In the latter, waves can be generated mostly in tropospheric regions by several sources among which the most relevant are flow over orography, convective clouds, and jets and fronts. After being emitted, they propagate vertically from the troposphere into the middle atmosphere. As the density decreases with the altitude, the amplitude of gravity waves grows until they break and eventually dissipate. Breaking process can cause turbulence and divergence of the momentum flux, and these can act as sources for other phenomena, for example clear air turbulence, quasi-biennal and semiannual oscillations, or for the generation of other gravity waves, called secondary waves. In addition, gravity waves are known to have a substantial impact also in the oceans, were they transport energy, and upon breaking contribute to the mixing (Sutherland et al. 2019). However, because the focus of this thesis is on atmospheric gravity waves, the following part of this introduction concentrates on the effects they have on atmospheric phenomena.

Since 1960, when Hines (1960) first attributed moving irregularities in the upper atmosphere to internal gravity waves, the dynamical interaction of these waves with other phenomena has been extensively proven. At lower altitudes, for example, gravity waves are involved in clear-air turbulence (CAT), which is of great relevance for air transport since it can cause serious aircraft incidents. One of the causes for such incidents is that CAT usually occurs in areas devoid of clouds (hence the name) and, therefore, cannot be seen and avoided by pilots. Several studies (Sharman et al. (2012) and references therein) have established that breaking gravity waves can trigger clear-air turbulence.

Furthermore, upward propagating gravity waves play a dominant role in the tropical stratosphere by driving the quasi-biennial oscillation (QBO) and the semiannual oscillation (SAO) of the zonal wind at equatorial latitudes. The QBO is the reverse of zonal winds from easterlies to westerlies (and vice-versa), which occurs at latitudes between 15°N and 15°S with a periodicity of approximately two years. The SAO has a periodicity of about six months and influences the upper stratosphere and lower mesosphere. It was already proposed in the '70s that these oscillations arise from a combined effect of the interaction between internal-gravity waves and the mean flow and equatorial waves (Lindzen and Holton 1968). There is a fairly good agreement of the proposed driving of QBO by gravity waves with observations (Ern et al. 2014), whereas the contribution and more details about the gravity wave drag is still not completely understood (Ern et al. 2015).

In the mesosphere, gravity waves become even more significant: by breaking wave processes the momentum flux can become divergent and, therefore, decelerate the zonal mean winds causing the reversal of the vertical shear and induce a summer-to-winter pole residual circulation. A sketch of such residual circulation is depicted in figure 1.1, where the mesospheric branch driven by gravity wave breaking is visible at the top. Because of this circulation, with rising motions in the summer hemisphere which lead to adiabatic cooling, and sinking ones in the winter hemisphere which lead to adiabatic heating, the resulting meridional temperature gradient is inverted compared to the one induced by solar radiation and observed at lower altitudes (Fritts and Alexander 2003). Indeed, the mesosphere at the summer pole is characterised by the temperature minimum, whereas the mesosphere at the winter pole is relatively warm.

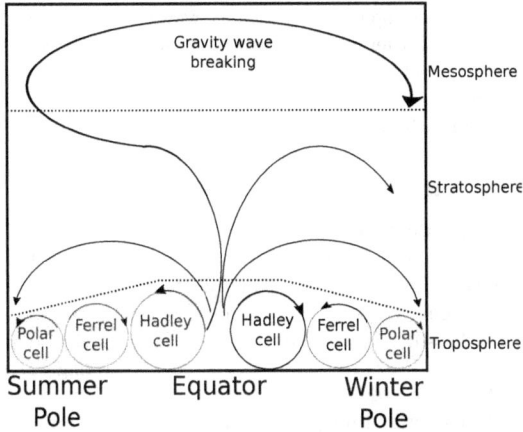

Figure 1.1: Schematic of the residual mean meridional circulation in the atmosphere. The circles denotes the Hadley, Ferrel, and Polar cells in the troposphere. The other lines indicated branches of the stratospheric and mesospheric circulation. The drawing is inspired by Plumb (2002).

It is clear that gravity waves have a substantial impact on several phenomena that are of relevance for the physics of the atmosphere; therefore, they cannot be neglected in realistic general circulation models and consequently in climate and weather models, which aim to predict the atmospheric dynamical features. Despite this, gravity waves have had a controversial position in the weather forecasts models. Even before such forecasts were run on computers, gravity waves had a fundamental role in the first ever prediction, done by hand by the father of the numerical weather forecasting Lewis Fry Richardson. As described in the book by Lynch (2006), Richardson's prediction of the change in the atmospheric pressure resulted in an unrealistic overestimation of two orders of magnitude because he used, as initial condition, the rate of pressure due to gravity waves to calculate a long term trend. To solve this problem and provide realistic forecasts several approaches have been followed, most of which are based on the same concept: the fast oscillation gravity waves need to be separated from the slowly evolving flow. One of the proposed and adopted solutions consists in the development of equations of motion that exclude gravity waves from the solutions. The development of models that completely filter out gravity waves—like the quasi-geostrophic model which eliminates the 'meteorological noises', as its founder Charney called gravity waves—is more than a simple matter of practicality. Indeed, this is at the core of the concepts of balanced and unbalanced dynamics and whether the evolution of the first one is possible in the complete absence of the second one.

Nowadays, it is clear that gravity waves cannot be filtered out of the numerical weather forecast and climate models, and that quasi-geostrophic equations are not suffi-cient enough to adequately describe the dynamics of the atmosphere, although they are useful for a basic understanding of the large-scale flow. However, the remaining problem is that part of the range of scales spanned by gravity waves is smaller than the current resolution of climate models and, therefore, need to be parametrised. Despite recent efforts to investigate IGW properties and their generation mechanisms, some aspects remain poorly understood, among which the gravity wave radiation process from jets and fronts which is an active research topic (Plougonven and Zhang 2014b).

1.1.1 General circulation

The Earth's atmospheric circulation is primarily driven by the incoming radiation from the Sun. Since the Earth's orbit around the Sun is elliptical and its axis has a current tilt of 23.4°, the energy received varies seasonally. To capture the main features of the observed atmospheric circulation in a simple model, we will neglect seasonal variations and assume that the tilt is 0°. In this configuration, the maximum incoming radiation is at the equator, and this leads to atmospheric motions acting to transport the heat excess from the equatorial towards the polar region. The resulting circulation would then be—similarly to what Hadley had imagined—two convective hemispheric cells extending from the equator to each of the poles, from where the warm air rises to where the cold air sinks. However, this single giant cell is not what is observed in the atmosphere for latitudes larger than 30° in both hemispheres, as it can be seen in the sketch in figure 1.2. Instead, three cells can be seen: two having the same overturning direction (polewards in the upper branch) and one in the opposite direction, sandwiched between them. Since there is a symmetry between the two hemispheres, we will focus our analysis on the North one. The winds at the surfaces, corresponding to the lower branch of each cell, are easterlies in the low latitudes. They then reverse to westerlies at mid-latitudes, and again easterlies (although weaker) near the poles.

To successfully model this observed global circulation of the Earth's atmosphere, a second forcing process has to be taken into account: the rotation around its axis. If friction is neglected, a parcel of air conserves the absolute angular momentum, $I = \Omega a^2 \cos^2 \phi + ua \cos \phi$ where Ω is the rotation, a, the radius, ϕ the latitude of the Earth, and u the eastward component of the wind. In the upper branch of the circulation cell, the air moving towards increasing latitudes will have an increasingly strong eastward wind whilst the opposite occurs in the lower branch, i.e. close to the surface where the so-called trade winds arise. The extension of the Hadley cell and the strength of the circulation can be obtained with a nearly inviscid theory by assuming an energetically closed circulation (Held and Hou 1980).

At mid-latitudes, the air is observed to circulate in the opposite direction, i.e. towards the equator, and the zonal symmetry of the flow breaks and waves appear. In this second cell, the 'Ferrel cell', the zonal winds become unstable—because of the 'baroclinic instability'—and show meanders of typical synoptical scales (Callies et al. 2014). The term synoptic derives from the Greek συνοπτικός, which literally means 'general view as a whole', but in meteorology it is used for large-space scales (Markowski and Richardson 2011), which have typical lengths of a few thousand kilometres. These winds and temperature fluctuations are associated with the formation of high- and low-pressure systems, which are mainly responsible for the weather variations at mid-latitudes. The effect of these eddies on the global circulation is to reduce the poleward temperature contrast by carrying cold air to the tropics and warm air towards the poles.

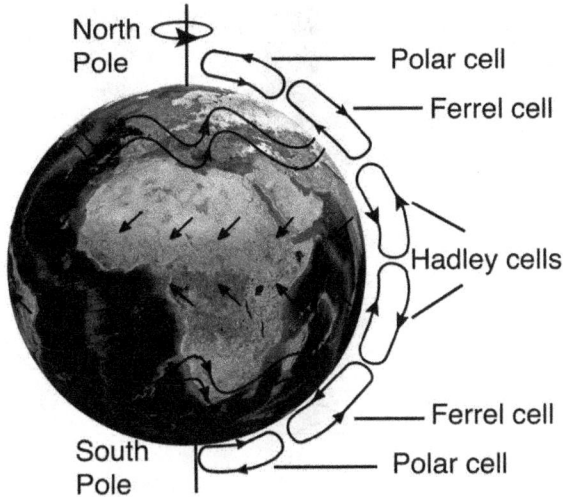

Figure 1.2: Schematic sketch showing some of the essential features of the atmospheric general circulation planet under equinox conditions. The three cells (Hadley, Ferrel, and polar) are drawn on the right hand side. The direction of the winds close to the surface are indicated on the globe by the black arrows and lines. At low latitudes, where the Hadley cells are developing, the winds are easterlies. For higher latitudes, under the Ferrel cells, westerlies with wavy patters can be observed. Near the poles, the winds turn again into easterlies (not indicated in the sketch). Drawing inspired by Wallace and Hobbs (2006).

The relevance of these large-scale waves to our study is that they have been observed to be a significant source of small-scale gravity waves (Ghil et al. 2010). The emission of gravity waves from jets and fronts is only one of the possible generation mechanisms. Other relevant sources in the atmosphere are the blowing of wind over topography and convective cumulus clouds. However, whilst these last two processes are reasonably well understood and flow-dependent parameterizations for the radiation of internal gravity waves from orographic and convective sources do exist, the situation is less developed for the gravity waves generated at the jets and fronts.

1.1.2 Gravity waves emitted from jets and fronts

Several atmospheric observations have highlighted consistent gravity wave activity at the jet front in the absence of orography. Uccellini and Koch (1987) firstly identified a common state of the background flow within which gravity waves are most likely to be emitted. In the studies considered, an enhancement of low-frequency gravity waves is recursively found in the exit region of the jet, where the flow decelerates. Successive observations (Fritts and Nastrom 1992, Sato 1994, Thomas et al. 1998, Hertzog et al. 2001, Wu and Zhang 2004) and numerical simulations (O'Sullivan and Dunkerton 1995, Zhang 2004) have confirmed these early studies and it is now accepted that gravity waves can be generated at the exit region of the jets. However, how these waves are generated is

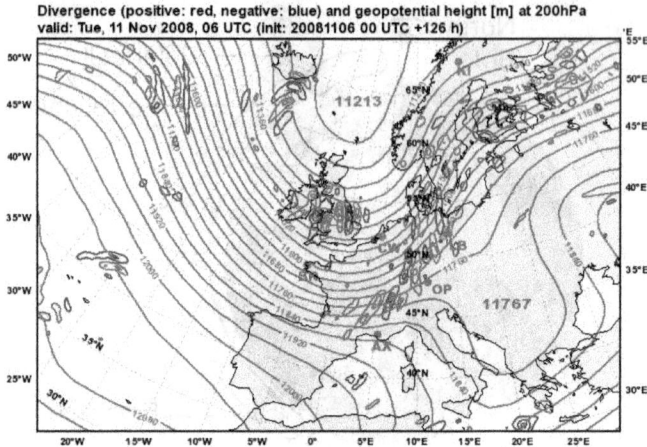

Figure 1.3: European Centre for Medium-Range Weather Forecasts (ECMWF) maps of horizontal divergence (negative values in blue, and positive values in red) at an altitude of 200 hP. The plot of the geopotential heights (green lines) show the ridges and troughs of the synoptic-scale waves.

still not completely understood. Furthermore, the exit region is also important for wave propagation e.g. via mechanisms like wave capture, which can modify the characteristics of the gravity waves moving through these areas (Bühler and McIntyre 2005, Plougonven and Snyder 2005). A horizontal map at 200 hPa showing the jet and the gravity wave packets along it is shown in figure 1.3 (the data and the plot are taken from the ECMWF model).

One of the main difficulties of studying gravity wave emission from jets and fronts is that the precise generation mechanism is still not fully evident since several potential sources coexist in the jet region, among which are the spontaneous imbalance of the mean flow, moist convection, and surface fronts (Wang et al. 2009). Moreover, contrarily to the case of orographic sources, there is no external forcing acting on the flow and, therefore, triggering the emission of gravity waves. In jets and fronts, the internal dynamic itself is responsible for such emission (Plougonven and Zhang 2014a). For this reason, one particularly challenging question is how the large scale flow, which is mostly balanced and only slowly evolving, can produce flow imbalance that forces the small-scales and high-frequency gravity waves. In fact, for many years scientists have debated whether this emission is even possible based on the existence or non-existence of the associated mathematical concept of the 'slow-mainfold'. According to Leith (1980) and Lorenz (1980) who postulated the existence of such a mainfold, balanced flows evolving along the slow mainfold will never emit fast waves since it would continue to remain balanced. The recent proof that such a slow mainfold cannot exist directly implies that balanced motions will inevitably emit internal-gravity waves and led to the introduction of the concept of spontaneous emission (Vanneste 2013). This represented a turning point in our understanding of source mechanisms in jets and fronts. Indeed, geostrophic adjustment was previously believed to be the most relevant source of gravity waves. This mechanism differs from spontaneous imbalance since it necessitates an initial imbalanced state of the flow, which emits gravity waves in its adjusting phase towards a balanced state. Spontaneous imbalance, on the other hand, assumes a completely balanced initial state.

One of the consequences of the non-existence of the slow-mainfold is that slow and fast motions are always coupled, and it is therefore impossible to obtain an exact separation of the two, although the interaction between them is usually weak. Indeed, the separation between the timescales of the balanced and unbalanced motions can be quantified with the Rossby number, defined as $Ro = U/fL$ where U and L are the typical velocity and scale of the flow, and f is the Coriolis parameter. The large-scale flow is characterised by $Ro << 1$ and it evolves at timescales longer than the inertial period. Gravity waves, however, have frequencies larger than f and, therefore, have timescales shorter than the inertial period. Nonetheless, balanced flows can locally exhibit Rossby number of order one or larger, and, in this case, the different scales are more strongly coupled.

Idealised simulations of vortex dipoles are often used to investigate the physical mechanisms of the spontaneous generation of gravity waves. The vortex dipole in a rotating, continuously stratified fluid is, for example, a simple model of a localised atmospheric jet which arises between a cyclone and an anticyclone propagating steadily for an extended period as a coherent structure. Studying the emission of gravity waves with such an approach has the advantage that the complexity of a time dependent and fast evolving background flow is eliminated and, therefore, it offers a simplified environment with respect to real flows.

The several numerical simulations done varying the numerical models and configurations (Snyder et al. 2007, Viúdez 2007, Wang et al. 2009) have consistently shown emission of gravity waves at the front of the dipole, which is identified with the exit region of the jet confirming what emerged from the atmospheric observations. The main features of the jet attributed to the emission of gravity waves are, on the one hand, strong velocities at the jet core producing strong advection, and on the other hand, significant variations along the jet which leads to the forcing (Plougonven and Zhang 2014b). Although these highly idealised models are useful to test theories and the validity of analytical models, they remain too far from the real flows in the atmosphere and have several drawbacks. Some of these are the systematic smaller amplitude of gravity waves generated in the models than the ones measured in the atmosphere and the suppression of other sources that can play a significant role in the jet and front systems.

1.2 Gravity waves in laboratory and numerical experiments

This thesis studies inertia-gravity waves emitted from jets and fronts by using laboratory experiments. Since these waves are characterised by low intrinsic frequencies, they are strongly influenced by the Earth's rotation. This investigation of the wave generation in real flows represents a fundamental bridge between the proposed theories and idealised experiments, such as jets within a vortex dipole, and the real atmosphere. Indeed, laboratory experiments have the advantage to provide a repeatable, more accessible and simplified version of the atmosphere, where some aspects, like topography, can be suppressed, but the complexity necessary to study nonlinear features in complex flows is conserved. By changing the experimental parameters, theoretical models and new hypothesis can be tested and validated. For these reasons, laboratory experiments offer great support in understanding the fundamental dynamical processes in the atmosphere.

Because of its proven capability to reproduce the main features of the large- and mesoscale motions in the atmosphere, namely large-scale circulation and baroclinic waves, the differentially heated rotating annulus experiment is chosen to conduct the investigations presented in this thesis. This experiment, introduced in the mid-1950s by Hide (1958) and Fultz et al. (1959), has been widely used over the years to study the manner in which the atmospheric circulation transports heat from equatorial to polar latitudes. The set-up of this experiment consists of a tank with three concentric cylinders filled with pure water. The inner cylinder (representing the polar region) is cooled, whereas the outer ring (representing the equator) is heated and the tank is mounted on a turntable, so it rotates around its vertical axis of symmetry. Therefore, the working fluid in the annular cavity is subject to a radial temperature difference and a Coriolis deflection. The combined effect leads, for high enough values of the rotation rate, to the baroclinic instability with the formation of cyclonic and anticyclonic eddies covering the full water column.

In recent years, this rotating annulus has been used as a testbed to study multiple-scale interactions both in numerical simulations (Jacoby et al. 2011, Randriamampianina and del Arco 2015, Von Larcher et al. 2018, Hien et al. 2018) and laboratory experiments (Vincze et al. 2016, Rodda et al. 2018). The numerical models, representing the differentially heated rotating annulus, have been used several times to investigate IGW generation mechanisms. Jacoby et al. (2011) identified short-period waves consistent with the theoretical dispersion relation of IGWs and indicated as a possible generation mechanism a boundary instability at the inner wall of the tank. IGWs have also been identified in numerical simulations in a version of a classical annulus set-up with continuous stratification by Randriamampianina and del Arco (2015). The characteristics of the obtained waves are similar to the ones found by Jacoby et al. (2011) although they suggested Kelvin-Helmholtz instability as the generating mechanism.

Other numerical investigations on the spontaneous emission of gravity waves from jet stream imbalance have been done by Borchert et al. (2014) using a finite volume model of the classic differentially heated rotating annulus. In their work, the numerical simulations have been conducted using two experiments; the first is a classical configuration of the rotating annulus with parameters very similar to those used in the laboratory experiment by Harlander et al. (2011), and a second using a wider and shallower configuration that is more atmosphere-like. The subsequent investigation conducted by Hien et al. (2018) with the atmosphere-like annulus showed that the balanced flow primarily forces the gravity waves, and distinguished then between the wave packets generated by the lateral wall boundary instability to the ones emitted by the flow. These numerical results, which first identified spontaneously emitted gravity waves in the differentially heated rotating annulus, encourage laboratory investigations with an analogue configuration for comparison, and to offer a validation for the simulations.

Experimentally, gravity waves have been observed in the differentially heated rotating annulus experiment by Read (1992), where the interaction with the temperature probes most likely caused the emission. Moreover, as mentioned above, IGWs excited by boundary layer instability have been found numerically and experimentally by Jacoby et al. (2011). So far, no evidence of gravity waves from jets and fronts can be found in the literature for this experimental laboratory set-up. In another experiment, i.e. the shear-driven version of the rotating annulus, internal gravity waves have been observed in the vicinity of the baroclinic jet by Lovegrove et al. (2000) and Williams et al. (2005). This experimental set-up differs from the differentially heated rotating annulus because the tank is filled with two immiscible fluids and the baroclinic instability is driven by the vertical shear created mechanically with a differentially rotating lid. With the help of a numerical model of the annulus, the generation mechanism of the observed gravity waves has been investigated and identified as spontaneous generation (Williams et al. 2008). A modified version of this two-layer experiment has been proposed by Flór et al. (2011), where they used salty stratified fluid instead of two fluids with different densities. The small-scale waves observed in this configuration are claimed to be generated by the Hölmböe instability occurring at the density interface.

In view of these results, one might wonder why internal gravity waves emitted in the vicinity of jets and fronts have not been observed experimentally in the differentially heated rotating annulus. One of the main problems encountered when using this experiment to investigate the emission and properties of internal-gravity waves is that its inverse aspect ratio ($\Pi = H/L$, the ratio between the vertical depth and the horizontal scale of motion) is usually greater than one. On the contrary, this ratio is much less than one in the atmosphere. For the differentially heated rotating annulus experiment, the buoyancy frequency can be expressed as $N = (g\alpha\Delta T/H)^{1/2}$, where g is the gravity acceleration, α the thermal expansion coefficient, ΔT the difference of temperature between the two walls, and H the total fluid depth. As a consequence of the large fluid depth, the narrow-gap classical configuration leads to a ratio between the buoyancy frequency N and the Coriolis frequency $f = 2\Omega$ (Ω being the angular frequency of the rotation of the tank) less than one, whilst in the atmosphere $N/f \sim 100$. This mismatch between the experiment and the atmosphere does not represent an impediment when investigating the large-scale dynamics, but it becomes fundamental for our studies of gravity waves. Indeed, the dispersion relation of inertia-gravity waves reads

$$\omega^2 = N^2 \cos^2(\alpha) + f^2 \sin^2(\alpha), \qquad (1.1)$$

where the intrinsic frequency ω of the wave is determined by the buoyancy frequency N, the Coriolis frequency f, and α, the angle between the phase velocity and the horizontal plane. Hence, IGW properties can be expected to differ between the experiment and the atmosphere. Considering (1.1) one can notice that the differences are not just quantitative, but also qualitative. In the atmosphere, the high-frequency waves propagate nearly horizontally (the low-frequency waves near vertically), whereas in this classic laboratory set-up the behaviour is just the opposite (Vincze and Jánosi 2016). For $N/f < 1$ waves can still propagate in the flow, but they are more like inertial waves, which are rotation dominated, and differ from gravity waves in several aspects. Since we aim to study atmospheric jet generated waves, which are gravity dominated, the only way to obtain similar features is to use a configuration of the experiment which allows $N/f > 1$. To reach this, two modified experiments are used for the work presented in this thesis.

One might be tempted to think that the problem could easily be solved by decreasing the fluid depth in the experiment so that N increases consequently and becomes bigger than f. However, two problems would then occur: the first one is that the fluid depth cannot be decreased arbitrarily since the friction at the bottom of the tank creates a boundary layer (the Ekman layer) and when its thickness reaches a specific ratio of the total fluid depth it will disturb the dynamics of the interior of the fluid finally preventing baroclinic waves from developing. The second point is that the baroclinic instability regime is reached for $NH/fL < 1$. For larger values, again the baroclinic waves do not occur in the fluid interior. For these reasons, it is not so trivial to modify the classical experiment in such a way that it becomes suitable for the study of gravity waves.

A first solution to increase N in the classical rotating annulus, without changing the experimental set-up, has been proposed by Vincze et al. (2016), who introduced a thermohaline version of the classical baroclinic wave tank experiment. In this modified set-up of the differentially heated rotating annulus, called the 'barostrat' experiment, a continuously stratified salinity profile is prepared in the experimental cavity before each measurement with the so-called double-bucket technique (Oster and Yamamoto 1963). By introducing a vertical salinity gradient to the set-up, the frequency ratios reach $2 < N/f < 6$, which are much smaller but at least qualitatively similar to the atmospheric case. This particular configuration is especially interesting for investigating the wave regimes that develop in the fluid depth as well as studying inertial Kelvin/baroclinic wave coupling and the coexistence of different baroclinic waves, and, more importantly for our purposes, to investigate the occurrence of inertia-gravity waves spontaneously emitted by the baroclinic wave.

A second way to increase N/f is by decreasing the aspect ratio so that the baroclinic instability regime can be reached for shallower fluid depths. To do so, a new wider annulus has been built at the BTU laboratories, with geometrical parameters similar to the ones proposed by Borchert et al. (2014). The advantage of this second solution is that a correspondent numerical set-up exists and the data sampled in the laboratory and from the numerics can be compared, which is not the case for the barostrat experiment. The numerical simulations with the *cyfloit* model are run at the Goethe University in Frankfurt (GUF) by Steffen Hien, who provided the data that has then been analysed and used for a comparison with the data collected at the BTU in the laboratory.

1.3 Outline

The two modified configurations of the differentially heated rotating annulus, i.e. the barostrat and the atmosphere-like experiment, described in the previous section, are used in this thesis to study the generation of gravity waves at the jets and fronts. To start with, for each configuration, the regime of the large-scale baroclinic waves developing in the annulus are studied and presented to provide a description of the background flow from which inertia-gravity waves (IGWs) are emitted. Subsequently, a signature of IGWs is sought in the vicinity of the jet. Afterwards, the properties of such waves, e.g. frequencies, wavelengths, propagation direction and speed, are investigated in detail. Finally, possible generation and propagation mechanisms fitting the features of the IGWs observed are proposed. These results are split into two chapters; one devoted to the barostrat experiment (chapter 6), of which the contents are mostly based on the published article (Rodda et al. 2018), and the second one to the atmosphere-like experiment (chapter 7), which is based on the submitted paper (Rodda et al. 2019) and on the one in preparation (Rodda and Harlander 2019). The theoretical description of the phenomena, experimental apparatuses, measurement techniques, data post-processing and analysis are all treated in separated chapters that are organised as explained in the following.

In chapter 2, the most relevant phenomena are described from a theoretical point of view. The contents shown are not original work, but mainly reported from textbooks, and they provide the reader with the background necessary to understand the features observed in the experiments. The first part focuses on the large-scale flows and the underlying mechanisms of formation of baroclinic waves, whilst the second part focuses on smaller scales and in particular on inertia-gravity waves. These are treated with linear theory and derived as a combination of gravity and inertial waves. Section 2.3 is devoted to the fundamental equations governing the motions in rotating and stratified fluids. Afterwards, the balanced motions (geostrophic and quasi-geostrophic) are treated. In section 2.4, the baroclinic instability is explored following the inviscid model by Eady. In section 2.5, the fundamental features of gravity waves are described in the simplified context of linear theory, and finally, in section 2.6 the sources and propagation mechanisms relevant for gravity waves are discussed.

The two experimental set-ups are described in chapter 3. To begin with, the origins of the differentially heated rotating annulus and its analogy with the atmosphere are introduced. The flow regimes developing in the annulus are illustrated together with the inviscid analytical model developed by O'Neil (1969) that we use to compare with instability diagrams obtained with our experiments. Afterwards, an experimental study of the limits of the ratio between the Ekman boundary layer and the total fluid is shown. This is offered as evidence to the need of a wider tank configuration. In the last two sections, the barostrat experimental configuration and the new atmosphere-like tank are illustrated. For the barostrat experiment, the focus is on how the salinity profile is achieved and the double-diffusive convection regime that sets in as a consequence.

In chapter 4, the measurement techniques used in both experiments to study the flow are explained in detail. Firstly, the two devices used to measure temperature are presented with their technical details. These comprise of an infrared camera, which measures the temperature at the water surface, and several temperature sensors that can be placed at different fluid depths, and therefore, give an insight of the temperature in the fluid's interior. The other quantity measured in the experiment is the velocity. This is done using the Particle Image Velocimetry technique (PIV), which allows us to calculate two of the velocity components by using a laser light sheet and a camera co-rotating with the tank. Since this technique needs a sophisticated post-processing procedure done using the free MATLAB toolbox UVmat, the main operations are explained in the chapter.

In chapter 5, we present the methods used to analyse the data collected from the various measurements. For the large-scale flow characteristics, we use two statistical methods, namely the harmonic analysis and empirical orthogonal functions. These two are then applied to the barostrat experimental data to study the dynamics developing at different fluid depths. Afterwards, the analysis used to investigate the properties of the internal gravity waves is introduced. The first goal is to separate the signal of the small-scale waves from the rest of the flow. Then the frequencies and wavelengths of the observed waves are captured in Fourier space by performing two-dimensional Fourier transforms. Finally, the method of decomposition of the energy spectra, which separates the balanced from the imbalanced flow components, is presented.

Chapter 6 is wholly devoted to the results obtained within the barostrat experiment. PIV velocity maps are used to describe the wavy flow pattern at different heights. Using a co-rotating laser and camera, the wave field is well resolved and different wave types can be found: baroclinic waves, Kelvin and Poincaré type waves. The signature of small-scale IGWs can also be observed attached to the baroclinic jet. The baroclinic waves occur at the thin convectively active layer at the surface and the bottom of the tank, though decoupled they show different manifestation of nonlinear interactions. The inertial Kelvin and Poincaré waves seem to be mechanically forced. The small-scale wave trains attached to the meandering jet point to an imbalance of the large-scale flow. For the first time, the simultaneous occurrence of different wave types is reported in detail for a differentially heated rotating annulus experiment.

The results obtained from experiments done with the new atmosphere-like tank are included in chapter 7. The first part (until section 7.2) focuses on the large-scale flow. The flow observed in the experiment is compared to the one obtained in the numerical simulations run at the GUF by Steffen Hien. The conditions for gravity waves are then analysed and compared. The last section of this first part shortly shows preliminary results when some modifications are added to the laboratory set-up. In the second part, gravity waves in the atmosphere-like tank are investigated. First of all, their properties are described together with their positioning and how they move with respect to the baroclinic jet. Subsequently, the possible source mechanisms are analysed. Finally, the energy spectra for the classic configuration, the barostrat modification of it, and the atmosphere-like tank are compared with atmospheric spectra.

In the concluding chapter 8, a discussion and summary of the work presented are given. Based on the results and questions raised, some suggestions for future work are included as final remarks.

Chapter 2

Theoretical background

"It doesn't matter how beautiful your theory is, it doesn't matter how smart you are. If it doesn't agree with experiment, it's wrong. In that simple statement is the key to science."

— Richard P. Feynman —

This chapter supplies the fundamental theory and equations to describe large-scale baroclinic waves and small-scale gravity waves. The aim is to provide the reader with the theoretical background necessary to understand the phenomena observed in the laboratory experiments, which are presented in the following chapters. A complete derivation of the equations is beyond the purpose of this thesis and, therefore, the reader is referred to textbooks (e.g. Sutherland (2010) and Nappo (2013) for gravity waves, and Pedlosky (1982), Vallis (2006), Cushman-Roisin and Beckers (2011) for the large scale flow).

In this thesis, we focus on phenomena occurring in the atmosphere, which similarly to the ocean, is a geophysical fluid. Geophysical flows exhibits peculiar properties because their motions are strongly influenced by the Earth's rotation and stratification due to changes in density and temperature.

2.1 Motion in a rotating fluid

The most natural choice for studying atmospheric motions is to describe them in a system of reference which rotates uniformly with planetary angular velocity Ω. Such rotating frame is a non-inertial frame, and therefore some additional fictitious forces are coming into play acting on the fluid motion (Pedlosky 1982). We can see them considering the transformation of the acceleration from an inertial frame to a rotating one, which reads

$$\left(\frac{d\vec{u}_I}{dt}\right)_I = \left(\frac{d\vec{u}_R}{dt}\right)_R + \underbrace{2\vec{\Omega} \times \vec{u}_R}_{\text{Coriolis}} + \underbrace{\vec{\Omega} \times (\vec{\Omega} \times \vec{r})}_{\text{centripetal}} + \underbrace{\frac{d\vec{\Omega}}{dt} \times \vec{r}}_{\text{rotation rate}}, \tag{2.1}$$

where the three additional terms on the right-hand side correspond to the Coriolis acceleration, the centripetal acceleration and the acceleration due to the variation in the rotation rate. The last term can be neglected for most of the atmospheric and oceanographic phenomena. The centripetal acceleration can be expressed as the gradient of a

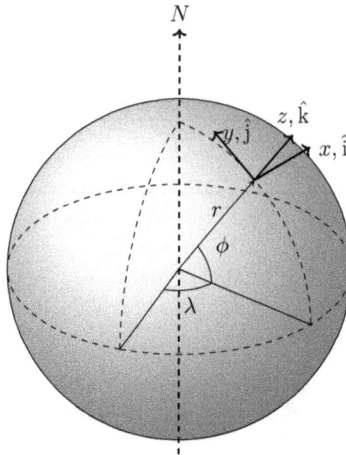

Figure 2.1: Local Cartesian coordinate system of reference on a spherical Earth. The coordinate x is directed eastward, y northward, and z upward.

potential and then, at least for homogeneous fluids, be added to the pressure term, which also includes the gravitational potential (note it is directed radially outwards). With these two approximations, the only additional term left in (2.1) is the Coriolis one. The direction of the Coriolis force is such that a moving fluid particle is deflected towards the right in the Northern hemisphere and towards the left in the Southern hemisphere (Holton 1973).

Because the Earth can be approximated as a spherical body, it might seem logical to use a spherical coordinate system to describe atmospheric and ocean motions. However, the sphericity of the Earth is not always important, especially for phenomena on a scale smaller than the Earth's radius for which it is more convenient to use a local Cartesian system, as the one shown in figure 2.1.

If we approximate the Earth to a perfect sphere rotating around its North-South pole axis, one can easily see from figure 2.1 that the rotation vector $\vec{\Omega}$ is parallel to the vertical direction z only at the poles. For all the other latitudes ϕ, it is convenient to decompose the vector into two components in the local cartesian system centred at a point at the latitude ϕ. The unit vectors describing these coordinates are $(\hat{\imath}, \hat{\jmath}, \hat{k})$ pointing eastward, northwards, and upwards respectively. The rotation vector can then be re-written as

$$\vec{\Omega} = \Omega \cos\phi\hat{\jmath} + \Omega \sin\phi\hat{k}. \tag{2.2}$$

The Coriolis term in equation 2.1 will then be

$$2\vec{\Omega} \times \vec{u}_R = (-2\Omega v \sin\phi + 2\Omega w \cos\phi)\hat{\imath} + (2\Omega u \sin\phi)\hat{\jmath}(-2\Omega u \cos\phi)\hat{k} \tag{2.3}$$

The coefficients in (2.3), can be defined as $f = 2\Omega \sin\phi$ and $\tilde{f} = 2\Omega \cos\phi$, where f is the Coriolis parameter, whereas $\tilde{f}$ is sometimes called reciprocal Coriolis parameter (Cushman-Roisin and Beckers 2011).

In the following, the **traditional approximation** will be used, according to which the terms $\tilde{f}$ are neglected. With this approximation, (2.3) reduces to

$$2\vec{\Omega} \times \vec{u}_R = (-2fv)\hat{i} + (2fu)\hat{j} \tag{2.4}$$

2.2 Motion in a stratified liquid

The temperature, pressure, and salinity (for the ocean), variations with depth and altitude are essential for the determination of density changes that result in the vertical structure of the atmosphere and the ocean (Sutherland 2010).

To better understand the effects of stratification on the fluid's motion, we start considering a fluid with density continuously decreasing with depth/height due to changes, in temperature, salinity, or both. If the fluid is inviscid and stationary, where Coriolis forces are absent, the static mechanical equilibrium is given by the balance between the gravity and the pressure forces

$$\frac{\partial p}{\partial z} = -\rho(z)g, \tag{2.5}$$

where the z-coordinate is taken to be upwards. The hydrostatic equilibrium states that the pressure (p) is constant at each horizontal level in the fluid.

Let us now consider a fluid parcel initially situated at the height z_0 and of density $\rho_0 = \rho(z_0)$. When the parcel is vertically displaced in the stably stratified fluid by δ_z, a restoring force is exerted on it and tends to bring it back at its initial position. Newton's law predicts the vertical motion of this fluid parcel

$$\rho_0 \frac{d^2\delta_z}{dt^2} = (\rho(z) - \rho_0)g. \tag{2.6}$$

If the displacement around the initial position is small, we can use a Taylor expansion around the initial position z_0 and rewrite

$$(\rho(z) - \rho_0)g \approx g\frac{d\rho}{dz}\bigg|_{z_0}\delta_z. \tag{2.7}$$

Substituting (2.7) into (2.6) we can derive the equation

$$\frac{d^2\delta_z}{dt^2} + N^2\delta_z = 0, \tag{2.8}$$

where

$$N^2 = -\frac{g}{\rho_0}\frac{d\rho}{dz}\bigg|_{z_0}, \tag{2.9}$$

is the squared Brunt-Väisälä or buoyancy frequency at which the fluid parcel oscillates.

Expression (2.8) is the equation of a harmonic oscillator the solutions of which are in the form $\delta_z(t) = A_0 sin(Nt + \phi)$ and N is the characteristic frequency of oscillation.

A plot of N^2 in the atmosphere is shown later in figure 2.9.

After discussing the effects of rotation and stratification, we now introduce some simplifying assumptions on the fluid behaviour that can be made for most of geophysical and laboratory systems. The model we heavily use is the

- **Boussinesq approximation**, which applies to processes that have vertical length scales smaller than the density variation scales. Under this approximation, all density fluctuations are neglected except when they appear in terms multiplied by the gravity force. This approximation includes the

- **Incompressibility approximation** which assumes that the velocities of the fluid are much smaller than the speed of the sound and therefore the fluid can be considered incompressible.

For large-scale flows we also consider the

- **Hydrostatic approximation** in the vertical momentum equation the gravitational term is assumed to be balanced by the pressure gradient term.

2.3   Fundamental equations

First of all, we want to derive the equations necessary to describe the transport of mass and heat in a fluid mathematically.

The fundamental governing equations of fluid dynamics are the mathematical statements of three fundamental physical principles upon which all of fluid dynamics is based, namely the mass conservation, momentum conservation, and the internal energy conservation.

Continuity equation is the relationship that expresses the conservation of mass in a fluid

$$\nabla \cdot \vec{u} = 0, \tag{2.10}$$

where $\vec{u}$ the velocity, and $\nabla = (\partial_x, \partial_y, \partial_z)$ is the nabla operator in the three-dimensional Cartesian coordinate system.

Momentum equation is the expression of Newton's second law for a fluid: the momentum density of a fluid parcel changes in time due to the sum of the forces acting on it. If we assume that the only real forces acting on a fluid are the pressure gradient force, gravitation, and friction, the acceleration of such a fluid, observed in a rotating frame of reference, is

$$\frac{D\vec{u}}{Dt} = \underbrace{-\frac{1}{\rho_0}\nabla p}_{\text{pressure gradient}} + \underbrace{-\frac{\rho}{\rho_0}\vec{g}}_{\text{buoyancy}} - \underbrace{(2\vec{\Omega}) \times \vec{u}}_{\text{Coriolis}} + \underbrace{\nu\nabla^2\vec{u}}_{\text{viscosity}}, \tag{2.11}$$

where the material derivative notation $D/Dt = (\partial/\partial t + \vec{u} \cdot \nabla)$ has been used. p is the pressure, $\vec{\Omega}$ is the angular velocity along the rotation axis, and ν is the kinematic viscosity. It should be noted that this simple form of the viscosity term is due to the fact that the flow is incompressible. The single forces acting to accelerate or decelerate the fluid parcel are indicated underneath the corresponding terms in the equation. The viscosity term can usually be neglected for atmospheric phenomena, although it can sometimes become important for laboratory experiments in some specific cases.

Thermodynamic energy equation follows from the conservation of energy if the system is adiabatic. In fluid dynamics, both liquids and gases are treated as fluids, and their governing equations are similar except for the equation of state.

For fluids, where $\rho = \rho_0[1 - \alpha(T - T_0)]$, the energy equation that governs the evolution of temperature is

$$\frac{DT}{Dt} = \kappa_T \nabla^2 T, \tag{2.12}$$

where T is the temperature of the fluid and κ_T is the kinematic diffusivity of the fluid.

The exact solution of these governing equations with proper initial and boundary conditions would give a complete description of the velocity, temperature, and pressure of the fluid. Unfortunately, due to the complexity of the equations, only rarely exact solutions can be found. We will now investigate which are the most important forces acting on the fluids and under which circumstances approximations can be made to further simplify the equations of motion.

2.3.1 Geostrophic and thermal wind balance

For geophysical fluids such as the atmosphere and the ocean, the Earth's rotation, appearing in the momentum equation thorough the Coriolis term, is much more relevant for large-scale slow flows than other terms, i.e. the viscous forces ($\nu \nabla^2 \vec{u}$) and the relative acceleration terms expressed by the material derivative $D\vec{u}/Dt$.

Dimensionless numbers, defined as ratios of the typical scales of different terms comparing in the equations of motion, can help in evaluating the importance of each term. For example, the ratio between the advective and the Coriolis terms defines the Rossby number

$$Ro = \frac{U}{fL}, \tag{2.13}$$

where $U = 10(0.1)$ m s^{-1} is the typical horizontal velocity in the atmosphere (ocean) and $L = 10^6(10^5)$ m is the typical horizontal scale. Since $f = 10^{-4}$ s^{-1}, we can immediately calculate $Ro = 0.1(0.01)$ for the turbulent atmosphere (ocean), showing that when the Rossby number is small, the advection terms are much less important than the Coriolis force. We can do a similar analysis for the viscous forces by defining the Ekman number

$$Ek = \frac{\nu}{\Omega H^2}, \tag{2.14}$$

where ν is the kinematic eddy viscosity and H the typical vertical scale of motion. Typical values for the atmosphere (oceans) are: $\nu = 5(10)$ m^2/s, $\Omega = 10^{-4}$ s^{-1}, and $H = 10^4$ m, which gives Ekman numbers of the order $10^{-3}(10^{-4})$. Therefore, when both the Ekman number and Rossby number are small, the Coriolis forces balance the pressure gradient force. The resulting motions are called geostrophic motions (from the Greek γη, which means Earth and στροφη, which means turning), and their governing equations reduce to

$$\vec{f} \times \vec{u} = -\frac{1}{\rho} \nabla_H p, \tag{2.15}$$

where $\nabla_H = \hat{i}\partial/\partial x + \hat{j}\partial/\partial y$ is the horizontal gradient operator and $f = 2\Omega\sin\phi$ is the local component of the planetary vorticity normal to the Earth's surface. From the geostrophic balance equation we can derive the geostrophic wind components $(\vec{u}_g = u_g\hat{i} + v_g\hat{j})$ in a cartesian coordinate system

$$u_g = -\frac{1}{\rho f}\frac{\partial p}{\partial y} \quad \text{and} \quad v_g = \frac{1}{\rho f}\frac{\partial p}{\partial x}. \tag{2.16}$$

These equations are valid for synoptic scale motions in the atmosphere and the ocean. One property of geostrophic fluids is that they flow along the isobars (lines of constant pressure), since the velocities are perpendicular to the horizontal pressure gradient, as it can be seen from equation (2.16). It can also be noted that since we dropped the time derivatives, the geostrophic equations are diagnostic, meaning that they cannot be used to predict the evolution of the velocity.

If the Coriolis parameter f is constant and the density is homogeneous

$$\nabla_H \cdot \vec{u}_g = 0, \tag{2.17}$$

The vertical derivative of the geostrophic winds can be calculated from (2.16) and by using the hydrostatic balance (2.5) it can be seen that they are identically zero

$$\frac{\partial u_g}{\partial z} = 0 \quad \text{and} \quad \frac{\partial v_g}{\partial z} = 0. \tag{2.18}$$

These equations, also known as the *Taylor-Proudmann theorem*, show that if the fluid is homogeneous and is confined by a horizontal rigid boundary, then the vertical derivative of the horizontal velocity components is equal to zero everywhere in the fluid, which implies that the flow moves along horizontal planes perpendicular to the rotation axis (Vallis 2006). Although the winds observed in the real atmosphere are not in perfect geostrophic balance, their motions are in near geostrophic balance so they can be considered as almost two dimensional flows.

As the Rossby number sizes the importance of rotation, the Froude number measures the influence of stratification

$$Fr = \frac{U}{NH}, \tag{2.19}$$

with the buoyancy frequency $N \sim 10^{-2}$ s^{-1}, the typical vertical scale $H \sim 10^3$ m. The typical velocities for the large-scale circulation are $U \sim 10$ m s^{-1} for the atmosphere and $U \sim 10^{-1}$ m s^{-1} for the oceans. These values give $Fr \sim 1$ for the first and $Fr \sim 10^{-2}$ for the latter.

For a rotating a stratified flows, the proportion of the two is given by the Burger number

$$Bu = \left(\frac{NH}{fL}\right)^2. \tag{2.20}$$

When $Bu \sim 1$, rotation and stratification assume the same importance in influencing the flow. The length scale of such a flow is

$$L_D = \frac{NH}{f}, \tag{2.21}$$

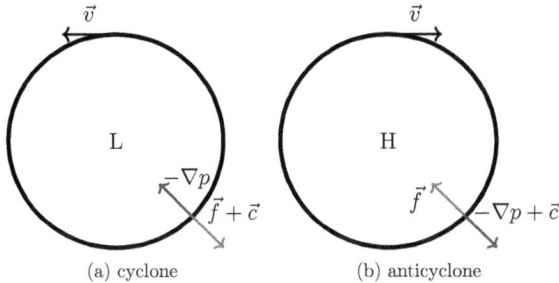

(a) cyclone   (b) anticyclone

Figure 2.2: Schematic representation of force balances in (a) a low, and (b) a high pressure system (in the Northern Hemisphere).

which is called Rossby radius of deformation.

If we now assume that the density depends on temperature and variates linearly in the form $\rho = \rho_0(1 + \alpha T)$, where α is the thermal expansion, by differentiating (2.16) with respect to the temperature and using the hydrostatic balance 2.5, we obtain the *thermal wind equation*

$$\frac{\partial u_g}{\partial z} = -\frac{\alpha g}{f}\frac{\partial T}{\partial y} \quad \text{and} \quad \frac{\partial v_g}{\partial z} = -\frac{\alpha g}{f}\frac{\partial T}{\partial x}. \tag{2.22}$$

This expression shows the relation between the geostrophic wind and the temperature. More specifically, it implies that a vertical gradient of the two wind components is associated with a horizontal gradient of the temperature. The thermal wind equation shows why the winds in the atmosphere increase with height.

For our discussion, it is useful to get a picture of the geostrophic balance for vortical flows in which the motion is axisymmetric. By using a cylindrical coordinate system, in the frame rotating with the vortex we can obtain a generalised expression of the geostrophic equation called the gradient wind equation

$$fv + \frac{v^2}{r} = -\frac{1}{\rho}\frac{\partial p}{\partial r}. \tag{2.23}$$

This equation expresses a balance between the centrifugal ($\vec{c} = v^2/r$) and Coriolis ($\vec{f} = fv$) forces and the radial pressure gradient, shown in figure 2.2 for a low (cyclone) and a high-pressure centre (anticyclone) in the Northern hemisphere.

2.3.2 Quasi-Geostrophic motion

Because, as we have seen, the geostrophic balance is a linear and diagnostic relationship, such equations are not sufficient to fully determine the dynamic of motions. To have prognostic equations, and therefore be able to predict the evolution of the flow in time, we have to consider higher-order dynamics in terms of small departures from geostrophic equilibrium. A detailed derivation of the equations can be found in textbooks (Vallis 2006), and it consists in expanding the non-dimensional variables in power series of the Rossby number $((\vec{u}, p) = (\vec{u}_g, p_g) + Ro(\vec{u}_a, p_a) + ...)$ and substituting them into the equations of motion, under the beta-plane approximation $f = f_0 + \beta y$, where $\beta = df/dy = 2\Omega \cos \phi / a$ is a positive constant with a the Earth's radius. This last approximation allows us to study the effects of the variation of the Coriolis parameter f with latitude, which are relevant for large-scale flows, by keep using the Cartesian coordinate system without need to switch to the more complicated spherical geometry.

At the zero-order, the dominant terms of the momentum equation (2.11) are the geostrophic wind components (2.16). At the first-order prognostic equations can be derived for the geostrophic wind. We here write the quasi-geostrophic equation in the form of the quasi-geostrophic potential vorticity Q (Vallis 2006)

$$\frac{\partial Q}{\partial t} + \vec{u} \cdot \nabla Q = 0, \tag{2.24}$$

where

$$Q = \nabla^2 \psi + \beta y + \frac{1}{\rho} \frac{\partial}{\partial z} \left(\rho \frac{f_0^2}{N^2} \frac{\partial \psi}{\partial z} \right). \tag{2.25}$$

The geostrophic streamfunction ψ is related to the velocity according to

$$u_g = -\frac{\partial \psi}{\partial y}, \quad v_g = \frac{\partial \psi}{\partial x}. \tag{2.26}$$

The three components of the quasi-geostrophic potential vorticity Q are: the geostrophic relative vorticity, the planetary vorticity, and the stretching vorticity. Because of the conservation of Q in flows where diabatic heating or frictional torques are absent (see Ertel's theorem in Vallis (2006)), the sum of its components terms is conserved following the geostrophic motion. The quasi-geostrophic equations describe the large-scale balanced motions, and are used in balanced models. These models are appropriate to capture the main characteristics of the large-scale flows, however, they filter out by construction the smaller-scale and rapidly variating internal gravity waves. Therefore, small-scale waves need to be treated differently, as we shall see later in section 2.5.

2.4 Baroclinic waves

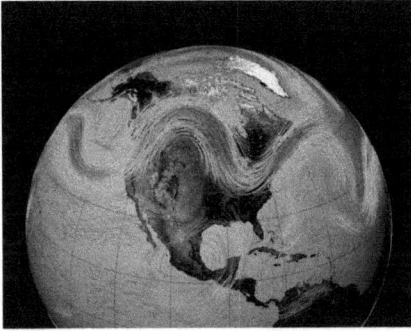

Figure 2.3: Schematic depiction of jet stream, with troughs (southward excursions) and ridges (northward excursions) that amplify in regimes of atmospheric blocking. Picture taken from NASA/GSFC (2012).

Mid-latitude weather variability is predominantly driven by extratropical cyclones and anticyclones systems. These motions are manifestations of a synoptic-scale instability of the atmospheric jet stream that can be seen in the weather maps (see figure 2.3) as wavelike horizontal excursions (of characteristic wavelengths between 3000 and 4500 km) of temperature and pressure contours, superposed on eastward mean flow. This hydrodynamic instability is called 'baroclinic instability', which if a combination of the words baro (pressure) and cline (slope), and it arises from sloped pressure contours. Such instability can be described as a form of 'sloping convection', and it is associated with vertical shear of the mean flow (Hide and Mason 1975). Baroclinic instabilities grow by converting potential energy related to the mean horizontal temperature gradient that, as we have seen in (2.22), arises to provide thermal wind balance for the vertical shear in the basic state flow (Holton 1973).

To illustrate the fundamental mechanism of baroclinic instability as a form of 'sloping convection' let us consider a stably stratified fluid in which density decreases with height and increases polewards. The isopycnals, i.e. the lines of equal density, will, therefore, be inclined by a certain angle with respect to horizontal surfaces. The combination of a heavier fluid at the bottom and variations of densities along horizontal surfaces is the condition for sloping convection to rise (when the fluid has lighter fluid at the bottom, but constant density along the horizontal, Rayleigh–Bérnard-type convection takes place instead).

Let us now have a look at the energy and consider the interchanging of two fluid particles in figure 2.4, where all the possible interactions with the rest of the fluid are neglected. The continuous sloping grey lines are isopycnals, and their tilt is affected by the rotation rate. When A is lifted upwards to replace C the restoring forces will move the particles (downwards and upwards respectively) back to the original positions since A is heavier than C. If, however, the parcel A is moved along a slant path and interchanged with B, because the latter is denser than A (even though it was initially above it) buoyancy will tend to separate the particles even further and therefore this situation leads to instability. In the latter case, therefore, the centre of gravity of the fluid has been lowered, and so its overall potential energy diminished. This loss in potential energy of the basic state must be accompanied by a gain in kinetic energy of the perturbation. Thus, the perturbation amplifies and converts potential energy to kinetic energy. The same would apply for any pair of fluid parcels lying between the geopotentials and isopycnals, in the

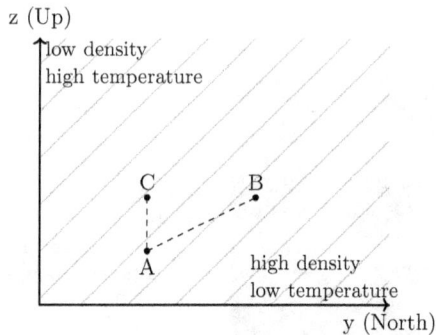

Figure 2.4: A sketch of the mechanism of the sloping convection. The grey lines are isopycnals (i.e. lines of equal density).

so-called 'wedge of instability'. In absence of rotation, an equator to pole overturning circulation would set in with the effect of flattering the isopycnals. If, as in the atmosphere, the density variations along level surfaces are maintained by differential heating, which is higher in the equatorial region and decreases towards the poles, this overturning circulation is menteined.

When rotation comes into play, we have seen that the thermal wind balance (2.22), geostrophy (2.16) and hydrostaticity (2.5) maintain the flow in equilibrium (Cushman-Roisin and Beckers 2011). Indeed, a poleward increase of temperature is balanced by westerlies winds increasing with heights without any motion in the meridional direction. However, we have previously seen that this is not the state of minimum energy. The release of energy for this flow occurs through the baroclinic instability which causes the formation of eddies. The dynamics of such mechanism is explained with the Eady linear model in the next section.

2.4.1 Inviscid theory–Eady model

The two earliest mathematical models of the baroclinic instability were developed independently in the late '40s by Charney (1947) and Eady (1949), and both show that the disturbances observed in mid-latitude in the atmosphere could be explained as a manifestation of such instability of the zonal winds (Pedlosky 1982). One of the main differences between the Eady and the Charney problem is that the latter is more complete and it includes the β-effect. The Eady problem, however, is clearer in its mathematical formulation and therefore easier to understand. The reader is referred to chapter 6 of Vallis (2006)) for a complete derivation of the equation shown in this section.

The Eady model is based on the following simplifying assumptions for a quasi-geostrophic and inviscid fluid:

1. the motion of the fluid is in an infinitely long channel of width L, confined in the vertical direction between two rigid, flat horizontal boundaries and the effect of friction is neglected;

2. the motion is on an f-plane, i.e. the effect of the Earth's sphericity is neglected;

3. the fluid is uniformly stratified ($N^2 = $ const);

4. the velocity is y independent and has a uniform vertical shear $U_0(z) = \Lambda z$, $\Lambda = $ constant.

5. only small amplitude waves are considered, so that the linear perturbation theory can be applied

The dispersion relation for Eady modes can be calculated from the quasi-geostrophic equation in the form of the potential vorticity (2.25) for a mean streamfunction of the form $\psi = -\Lambda z y$ (see chapter 6.6 in Vallis (2006)). By assuming conservation of the potential vorticity, applying the appropriate boundary conditions, and seeking for solutions in the form of linear waves, the dispersion relation reads

$$c = \frac{\Lambda H}{2} \pm \frac{\Lambda H}{\mu} \left[\left(\frac{\mu}{2} - \coth \frac{\mu}{2} \right) \left(\frac{\mu}{2} - \tanh \frac{\mu}{2} \right) \right]^{1/2}. \tag{2.27}$$

where $c = \omega/k$ is the phase speed, $\mu^2 = L_D^2(k^2 + l^2)$ is a horizontal wavenumber of zonal k and meridional l components scaled by the Rossby radius of deformation L_D defined in (2.21).

Unstable solutions i.e. solutions with amplitudes that grow exponentially with time at a rate $\exp(\sigma t)$, can be found when the phase speed $c = c_i$ is imaginary, since the waves are proportional to $\exp(-ikct)$. The solution with the largest magnitude of imaginary phase speed, will be the one that grows fastest and will be the dominant in the system.

The condition for an instability, therefore, is

$$\frac{\mu}{2} < \coth \frac{\mu}{2}, \tag{2.28}$$

which is satisfied when $\mu < 2.399$.

The growth rate is $\sigma = kc_i$

$$\sigma = k \frac{\Lambda H}{\mu} \left[\left(\frac{\mu}{2} - \coth \frac{\mu}{2} \right) \left(\frac{\mu}{2} - \tanh \frac{\mu}{2} \right) \right]^{1/2}. \tag{2.29}$$

It can be seen that for a given zonal wavenumber k the most unstable growing mode is the one for which the meridional wavenumber $l = 0$. Moreover, the maximum growth rate in (2.29) is obtained for $k = 1.61/L_D$. The growth rate as a function of the zonal wavenumber is plotted in figure 2.5, where the black vertical line indicates the maximum for a unitary Rossby deformation radius ($L_d = 1$). If typical values for the atmosphere are chosen ($N \sim 10^{-2}$ s^{-1}, $H = 10$ km, and $f = 10^{-4}$ s^{-1}), the Rossby deformation radius $L_d \sim 1000$ km and the corresponding horizontal wavelength ($\lambda = 2\pi/k$) is around 4000 km, which corresponds to the characteristic length scales of the large-scale waves observed in the atmosphere.

The vertical structure of the Eady wave is given by

$$\psi(z) = A \cosh \left(\frac{\mu}{H} z \right) - \frac{\Lambda H}{\mu c} A \sinh \left(\frac{\mu}{H} z \right), \tag{2.30}$$

(a)   (b)

(c)

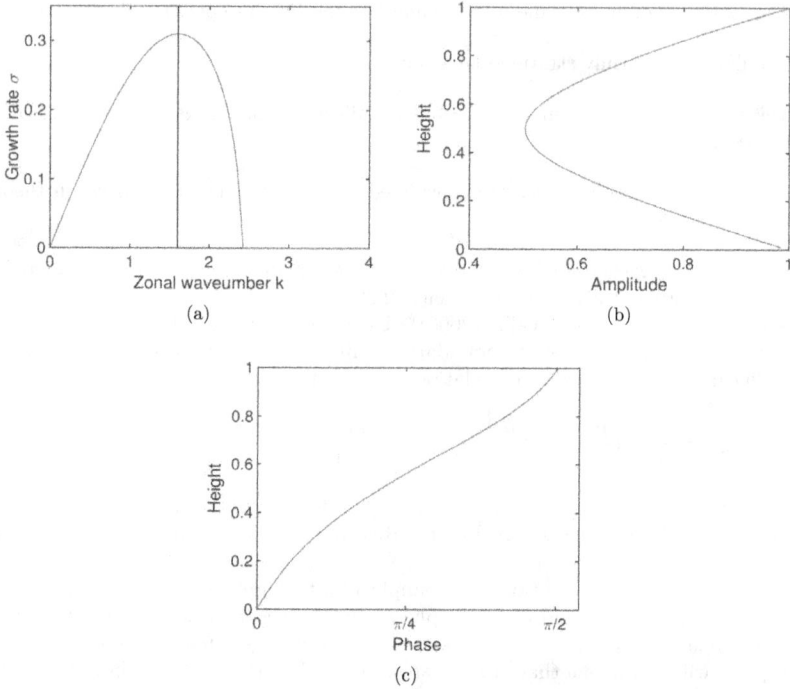

Figure 2.5: (a) Growth rate σ as a function of the zonal wavenumber k calculated from (2.29), (b) variations of the amplitude $|\psi|$ calculated with (2.31) with height, and (c) variations of the phase ϕ calculated with (2.32) with the height. The black vertical line in (a) marks the maximum growth rate at $k = 1.61$ with $L_D = 1$.

where $A = $ constant is an arbitrary factor. It follows that the wave amplitude variation with height is

$$|\psi(z)| = \left[\left(\cosh(\mu z/H) - \frac{c_r \sinh(\mu z/H)}{\mu|c|^2} \right)^2 + \frac{c_i^2 \sinh^2(\mu z/H)}{\mu^2|c|^4} \right]^{(1/2)}, \tag{2.31}$$

whereas the phase angle is

$$\phi(z) = \tan^{-1} \left(\frac{c_i \sinh(\mu z/H)}{\mu|c|^2 \cosh(\mu z/H) - c_r \sinh(\mu z/H)} \right). \tag{2.32}$$

The plot of the normalised (where $\Lambda = H = 1$) amplitude $|\psi(z)|$ is visible in figure 2.5 (b). What can be noted in the plot, is that the amplitude has two maxima at the top and bottom boundaries, and then it decreases towards mid depths. Indeed, two distinct Eady modes form at the rigid walls and exponentially decay towards the fluid interior.

When these two waves are mutually interacting then the instability can occur. For this to happen, the two waves have to be out of phase, and this condition is satisfied when $\phi = \pi/2$ (see figure 2.31(c)), which translates in a westward tilt of the wave with height. It is because of this tilt, combined with geostrophic and hydrostatic balance, that the release of potential energy and the poleward transport of warm fluid can be explained (Vallis 2006).

2.5   Internal gravity waves—theory

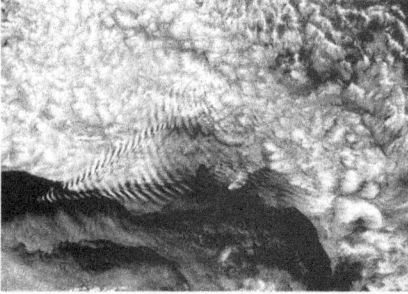

Figure 2.6: Satellite image from NASA Earth
Observatory (2005) of atmospheric internal
waves behind the Amsterdam island located
in the south of Indian Ocean.

Internal gravity waves occur in the interior of stably stratified fluids, and therefore
they naturally exist in the atmosphere and
oceans.

As for the surface gravity waves, buoyancy provides the restoring force that opposes vertical displacement from a stable
position. However, differently from surface waves, which are confined at the interface between the water and the air, internal gravity waves move within the fluid
and can, therefore, propagate vertically as
well as horizontally through it.

Besides stratification, Earth's rotation also
plays an important role for internal gravity
waves with periods comparable to a day.
Indeed in the atmosphere, gravity waves
have intrinsic frequencies variating over a
broad range, and only for high frequency waves the effects of rotation can be neglected, whereas for middle and low frequency waves rotation has an important influence.
The latter are then called 'inertia-gravity waves'.

Although the nature of gravity waves observed in the atmosphere (and in the laboratory
experiment) is nonlinear and waves interact with each other and with the mean flow, we
will here discuss the waves properties only by mean of linear theory.

As already done in the previous sections, we consider an incompressible, stratified,
rotating, inviscid, and adiabatic fluid. The deriving set of equations which we shall
consider in the following sections is the Boussinesq equations

$$\frac{\partial u}{\partial t} - fv = -\frac{1}{\rho_0}\frac{\partial p}{\partial x}, \tag{2.33a}$$

$$\frac{\partial v}{\partial t} + fu = -\frac{1}{\rho_0}\frac{\partial p}{\partial y}, \tag{2.33b}$$

$$\frac{\partial w}{\partial t} = -\frac{1}{\rho_0}\frac{\partial p}{\partial z} - \rho g, \tag{2.33c}$$

$$\frac{\partial u}{\partial x} + \frac{\partial v}{\partial y} + \frac{\partial w}{\partial z} = 0, \tag{2.33d}$$

$$\frac{\partial \rho}{\partial t} - \frac{\rho_0}{g}N_0^2 w = 0, \tag{2.33e}$$

where $N_0 = -g/\rho_0\partial_z\rho_0$ is the buoyancy frequency, which is constant if we consider a
uniformly stratified fluid [1].

---

[1]It can be shown that this condition is not restrictive as even for the case of a varying background
density $\rho(z)$ the derivative $\partial\rho_0/\partial z$ can be neglected in the governing equation for density (Pedlosky 2013).

We will now study three cases: purely gravity waves, where we neglect rotation, purely inertial waves, where stratification is not considered, and finally inertia-gravity waves, for which both rotation and stratification are important.

2.5.0.1 Linear equations for purely gravity waves

We start by considering the non-rotational case. The linearised equations of motion for this fluid become then

$$\rho_0 \frac{\partial u}{\partial t} = -\frac{\partial p}{\partial x}, \tag{2.34a}$$

$$\rho_0 \frac{\partial v}{\partial t} = -\frac{\partial p}{\partial y}, \tag{2.34b}$$

$$\rho_0 \frac{\partial w}{\partial t} = -\frac{\partial p}{\partial z} - \rho g, \tag{2.34c}$$

$$\frac{\partial u}{\partial x} + \frac{\partial v}{\partial y} + \frac{\partial w}{\partial z} = 0, \tag{2.34d}$$

$$\frac{\partial \rho}{\partial t} - \frac{\rho_0}{g} N_0^2 w = 0. \tag{2.34e}$$

This can be re-arrange the system of equations into a matrix (Sutherland 2010)

$$\begin{bmatrix} \rho_0 \partial_t & 0 & 0 & 0 & \partial_x \\ 0 & \rho_0 \partial_t & 0 & 0 & \partial_y \\ 0 & 0 & \rho_0 \partial_t & g & \partial_z \\ \partial_x & \partial_y & \partial_z & 0 & 0 \\ 0 & 0 & -N_0^2 \rho_0/g & \partial_t & 0 \end{bmatrix} \begin{bmatrix} u \\ v \\ w \\ \rho \\ p \end{bmatrix} = 0.$$

By taking the determinant of the differential matrix operator, we can get a single partial differential equation for one variable (see appendix A.1 for more details). For example for w the equation is

$$\frac{\partial^2}{\partial t^2} (\nabla^2 w) + N_0^2 \nabla_h^2 w = 0. \tag{2.35}$$

Note that the term multiplying N_0^2 (we shall drop the subscript from now on for simplicity) involves only horizontal derivative, pointing out that horizontal and vertical directions have dynamically different significance in a stratified flow (Pedlosky 2013).

We now make the Ansatz that the solutions are in the plane-wave form

$$w = \Re\{W_0 \exp[i(\vec{k} \cdot \vec{x} - \omega t)]\}, \tag{2.36}$$

where W_0 is the amplitude of the wave that here we assume to be constant and $\vec{k} = (k, l, m)$ is the wave vector, with k, l, m the wavenumbers along the x, y, and z coordinates respectively, and ω is the frequency of the wave.

We can solve (2.35) for (2.36) and this gives the dispersion relation for internal waves (see appendix A.2)

$$\omega^2 = \frac{N^2 K_H^2}{K^2}, \tag{2.37}$$

where $K^2 = k^2 + l^2 + m^2$, and $K_H^2 = k^2 + l^2$.

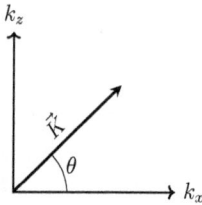

Another way to write the dispersion relation is in polar coordinate where $K_H/K = \cos\theta$. θ is the angle formed by the wave vector to the horizontal in the $\vec{k}$-space, as it can be seen in figure 2.7. The dispersion in this coordinate system becomes

$$\omega^2 = N^2 \cos^2\theta. \tag{2.38}$$

Figure 2.7

From the second expression, one can see that the frequency depends only on the orientation of the wave vector and not its magnitude. In other words, the phase lines are oriented with an angle that depends only upon the frequency of the waves and not on their spatial scale. This is a peculiar feature of gravity waves (Sutherland 2010).

Phase and group velocity The phase speed is defined as the speed the wave crests move.

$$\vec{c}_p = \frac{\omega}{K^2}\vec{k} \tag{2.39}$$

Using the two expressions found for the dispersion relation (2.37) (2.38) we can write ω explicitly as

$$\vec{c}_p = \frac{NK_H}{K}\frac{\vec{k}}{K^2} = N\cos\theta\frac{\vec{k}}{K^2}. \tag{2.40}$$

The group velocity is the velocity at which energy propagates and corresponds to the speed at which the wave packet travels.

$$\vec{c}_g = \nabla_{\vec{k}}\omega = \left(\frac{\partial\omega}{\partial k}, \frac{\partial\omega}{\partial l}, \frac{\partial\omega}{\partial m}\right). \tag{2.41}$$

As previously done for the phase velocity, we can rewrite the group velocity by using the dispersion relation in the form

$$\vec{c}_g = \left(\frac{Nm^2k}{K^3K_H}, \frac{Nm^2l}{K^3K_H}, -\frac{NmK_H}{K^3}\right). \tag{2.42}$$

It is easy to verify that

$$\vec{k}\cdot\vec{c}_g = 0, \qquad \vec{k}\times\vec{c}_p = 0. \tag{2.43}$$

These relations show that the group velocity is perpendicular to the wave vector and the phase velocity is parallel to it. Therefore, the energy propagates at 90° with respect to the crests.

2.5.0.2 Linear equations for purely inertial waves

We follow the same procedure used for purely gravity waves, but this time assuming a fluid which is homogeneous, inviscid, adiabatic, and with no mean background flow. In this case, the linearised equations of motion are

$$\rho_0 \frac{\partial u}{\partial t} - \rho_0 f v = -\frac{\partial p}{\partial x}, \tag{2.44a}$$

$$\rho_0 \frac{\partial v}{\partial t} + \rho_0 f u = -\frac{\partial p}{\partial y}, \tag{2.44b}$$

$$\rho_0 \frac{\partial w}{\partial t} = -\frac{\partial p}{\partial z}, \tag{2.44c}$$

$$\frac{\partial u}{\partial x} + \frac{\partial v}{\partial y} + \frac{\partial w}{\partial z} = 0, \tag{2.44d}$$

where $f = 2\Omega$ is the Coriolis parameter.

The equations written in matrix form are

$$\begin{bmatrix} \rho_0 \partial_t & -\rho_0 f & 0 & \partial_x \\ \rho_0 f & \rho_0 \partial_t & 0 & \partial_y \\ 0 & 0 & \rho_0 \partial_t & \partial_z \\ \partial_x & \partial_y & \partial_x & 0 \end{bmatrix} \begin{bmatrix} u \\ v \\ w \\ p \end{bmatrix} = 0.$$

and by taking the determinant, the resulting differential equation for the variable w is

$$\frac{\partial^2}{\partial t^2}\left(\nabla^2 w\right) + f^2 \frac{\partial^2}{\partial z^2} w = 0. \tag{2.45}$$

The wavelike solution gives the dispersion relation for inertial waves

$$\omega^2 = f^2 \frac{m^2}{K^2}. \tag{2.46}$$

or, in polar coordinates

$$\omega^2 = f^2 \sin^2 \theta. \tag{2.47}$$

where θ is, as for the gravity waves, the angle formed by the wave vector to the horizontal axis.

Phase and group velocity   The phase speed is defined as the speed the wave crests move.

$$\vec{c}_p = \frac{\omega}{K^2}\vec{k} = \frac{fm}{K}\frac{\vec{k}}{K^2} \tag{2.48}$$

The group velocity is

$$\vec{c}_g = \nabla_{\vec{k}}\omega = \left(\frac{\partial \omega}{\partial k}, \frac{\partial \omega}{\partial l}, \frac{\partial \omega}{\partial m}\right) = \left(\frac{-2\Omega km}{K^{3/2}}, \frac{-2\Omega lm}{K^{3/2}}, \frac{2\Omega K_H^2}{K^{3/2}}\right). \tag{2.49}$$

One of the characteristic properties of the inertial wave is that $\vec{c}_p \cdot \vec{c}_g = 0$, i.e. the group velocity is perpendicular to the phase velocity.

2.5.0.3   Linear equations for inertial-gravity waves

When the flow is both inertially and gravitationally stable, parcel displacements are resisted by both rotation and buoyancy. The resulting oscillations, a combination of the two aforementioned wave types, are the so-called inertia-gravity waves. For these waves, both Coriolis and gravity forces are affecting the propagation. The same derivation can be done for inertia-gravity waves and the matrix form of the governing equations is

$$
\begin{bmatrix}
\rho_0 \partial_t & -\rho_0 f & 0 & 0 & \partial_x \\
\rho_0 f & \rho_0 \partial_t & 0 & 0 & \partial_y \\
0 & 0 & \rho_0 \partial_t & g & \partial_z \\
\partial_x & \partial_y & \partial_x & 0 & 0 \\
0 & 0 & -N_0^2 \rho_0/g & \partial_t & 0
\end{bmatrix}
\begin{bmatrix}
u \\ v \\ w \\ \rho \\ p
\end{bmatrix} = 0.
$$

The dispersion relation is then

$$
\omega^2 = N_0^2 \frac{k^2 + l^2}{K^2} + f^2 \frac{m^2}{K^2}, \tag{2.50}
$$

or, in polar coordinates

$$
\omega^2 = N_0^2 \cos^2 \theta + f^2 \sin^2 \theta. \tag{2.51}
$$

where θ is the angle between lines of constant phase and the horizontal axis x. From the dispersion relation follows that high-frequency inertia-gravity waves ($\omega \simeq N$) have an angle with the horizontal axis $\theta \simeq 0$ and therefore propagates mostly vertically whereas low-frequency waves ($\omega \simeq f$) have $\theta \simeq 90°$ and propagate mostly horizontally.

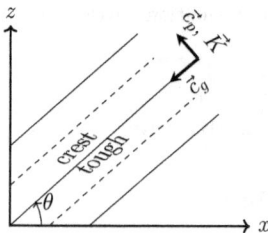

Figure 2.8

The schematic 2-dimensional draw in figure 2.8 shows the crests and toughs for IGWs and the directions of the wave vector $\vec{K}$, of the group velocity $\vec{c}_g$, and the phase velocity $\vec{c}_p$ with respect to them. As we have already seen in the cases of purely inertial waves and purely gravity waves, also inertia-gravity waves have group velocity perpendicular to the phase velocity ($\vec{c}_p \cdot \vec{c}_g = 0$) and the wave vector.

2.6   Internal gravity waves—sources and propagation mechanisms

Observations over the past years have established the significant role played by gravity waves in the global circulation of the atmosphere. These small-scale waves are mainly generated through various mechanisms in the troposphere and then, by propagating upwards through the stratosphere, they transport energy and deposit it in the mesosphere and even thermosphere when they break. The large amount of transported energy strongly influences the spatial and temporal characteristics of the middle and upper atmosphere (Nappo 2013). We will now get an overview of the principal sources and propagation mechanism of gravity waves in the atmosphere.

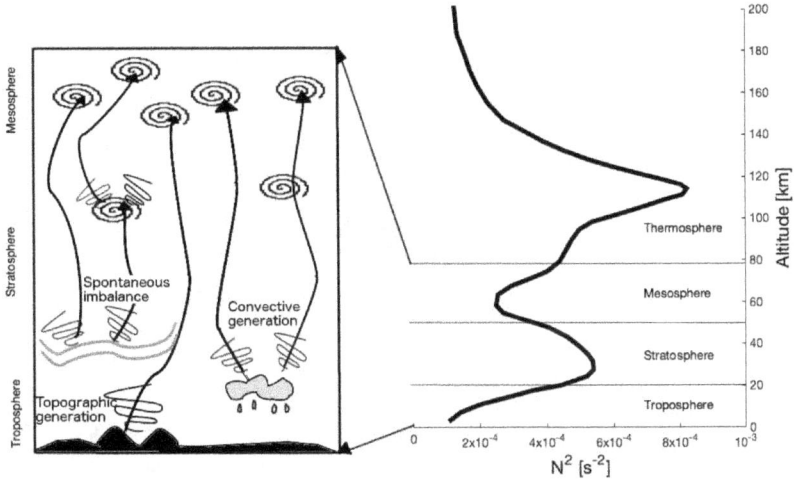

Figure 2.9: Schematics illustration of the processes that generate internal waves and ulti-mately lead to breaking in the atmosphere (adapted from Kim et al. (2003) and Sutherland et al. (2019)). The three main IGW sources are shown: generation from flow over topog-raphy, generation by convection, and spontaneous imbalance from the jet stream. The spiral symbol indicates wave breaking, which mainly occurs in the upper part of the at-mosphere. On the right-hand side the profile of N^2 as a function of the altitude is shown (adapted from Nappo (2013)).

2.6.1   Internal gravity waves sources

So far, we have discussed the conditions upon which internal gravity waves exist and propagate in a fluid without investigating the mechanisms that can trigger these waves from a fluid in a state of rest.

In the atmosphere, the most significant forcing mechanisms for internal gravity waves are orography, convection, and jet/front systems (a sketch is shown in figure 2.9). The distribution of non-orographic wave sources is clearly more spread out with respect to the orographic ones, which are tied to orographic features. For the first, convective sources are dominating at the Tropics while jets and fronts are the main sources at midlatitudes.

2.6.1.1   Topography

In nature, gravity waves can be triggered by mountain ridges, but also by isolated peaks, as in the case of volcanos (see i.e. picture 2.6). To get a basic understanding of the generation of waves from a flow over topography, we show a simplified problem for which solutions can be easily found. Many examples are proposed in the literature, perhaps the

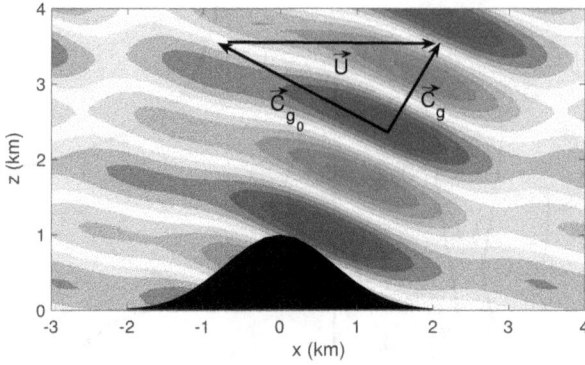

Figure 2.10: Stationary 2D flow over a gaussian-shaped hill. The colour shows the vertical velocity field (red upward and blue downward) obtained from (2.56).

most simple one is considering the flow over a periodic orographic terrain (Sutherland 2010).

Single three dimensional obstacles can be used to study the problem in laboratory experiments (Miles and Huppert 1968, Dalziel et al. 2011). We consider here a 2D uniform flow, with constant background stratification, over an isolated mountain of Gaussian shape to get some insights of this generating mechanism.

The surface height can be written as

$$h(x) = \exp(-bx^2). \tag{2.52}$$

The hight of the topography, here set to 1 km, is an essential parameter since it determines the amplitude of the generated IGWs.

When studying the generation of waves over orography, a fundamental parameter for determining the properties of the waves is the Scorer parameter S, named after the British meteorologist who studied mountain waves (Scorer 1949). The parameter is defined as

$$S = \frac{N}{U}. \tag{2.53}$$

Since usually in the atmosphere the buoyancy frequency N and the wind speed U are functions of the height z, also the Scorer parameter is height dependent. The equation for the stationary flow over the mountain can be written

$$\nabla^2 \psi + S^2 \psi = 0, \tag{2.54}$$

where ψ is the two-dimensional stream function

$$u = -\frac{\partial \psi}{\partial z}, \qquad w = \frac{\partial \psi}{\partial x}. \tag{2.55}$$

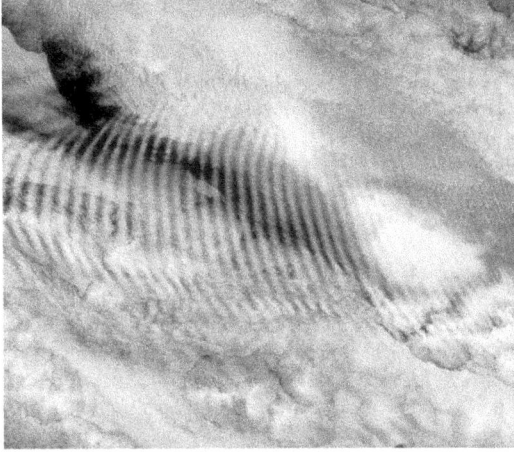

Figure 2.11: Gravity waves generated by rapidly rising deep convection over the ocean (picture from NASA Earth Observatory (2009)).

The problem can be solved by assuming linear wave solution of the form $\psi = \tilde{\psi}(x)\exp(ikx)$ and applying the linearised bottom boundary conditions $\psi = U_0 h(x)$. The vertical velocity, plotted in figure 2.10 is

$$w = -\frac{kU}{(b\pi)^{1/2}} \sum_{k=1}^{J} \exp\left(-\frac{k^2}{4b}\right) \sin\left(kx + (S^2 - k^2)^{1/2}z\right). \tag{2.56}$$

We can see from the group velocity c_g that the energy transport is vertical and downwind. Since the waves are stationary

$$\omega = 0, \quad c_p = 0 \tag{2.57}$$

Therefore, the intrinsic frequency, written as $\omega_i^2 = (\omega - Uk)^2$ becomes

$$\omega_i^2 = (Uk)^2 \tag{2.58}$$

2.6.1.2   Convection

One of the most important non-orographic source of IGWs is convection, particularly in regions located above the equatorial oceans and in the southern mid-latitudes in summer. Convectively generated gravity waves are also significant because they drive tropical circulation features like the equatorial stratospheric semi-annual oscillation (SAO) and quasi-biennial oscillation (QBO).

There are currently three convection excitation mechanisms proposed in the literature. The first involves vertical motions within storm clouds, where a time-varying thermal forcing is induced by latent heat release; the second consists of flow over the cloud tops or other obstacles produced by convective heating in clouds; the third is the vertical oscillations of the cloud tops. Moreover, wave generation is influenced by the collective behaviour of merging convective cells, frontogenesis and the thermodynamics of moist convection (Sutherland et al. 2019). Waves generated by convection are not characterised by a single prominent phase speed or frequency, as is the case for topographic waves but instead, they cover the full range of phase speeds, wave frequencies, and vertical and horizontal scales. The low-frequency waves, in particular, may be observed in the middle atmosphere at large horizontal distances from the convective source, making a correlation with clouds or other indicators of convection more difficult. In the tropics, however, far from topography and regions of baroclinic instability, the occurrence of inertia-gravity waves has been linked to convection as the source (Fritts and Alexander 2003).

One example of experiments investigating the propagation of convectively generated gravity waves in a stably stratified layer is presented by Le Bars et al. (2015).

On a more basic and practical point of view, we can consider the convective stability of the flow. We follow the method reported by Nappo (2013) which is based on stability analysis that consists on calculating the total energy variation of a system. More in detail, we can consider a system made of two parcels placed one on top of the other (at positions z_a and z_b respectively) and we look if the total energy increases or decreases upon the position swap of the two parcels. A system is unstable if the total energy decreases after the adiabatic exchange of positions. If the conditions of the fluid at the two heights are constant densities ρ_a and ρ_b and uniform velocities u_a and u_b, the total energy variation, under the assumptions of mass and momentum conservation, will be

$$\Delta E = g(\rho_a - \rho_b)(z_b - z_a) - \frac{1}{2}\frac{\rho_a \rho_b}{(\rho_a + \rho_b)}(u_a - u_b)^2. \tag{2.59}$$

The first term on the right-hand side is the difference between the final and the initial potential energy and the second term the difference of the final and initial kinetic energy. It follows that if $\Delta E > 0$ the flow is stable vice-versa.

Another way to express the instability of the flow is by using the *Richardson number*, which is a fundamental parameter used in turbulence theory and is defined as

$$Ri = \frac{N^2}{(\frac{dU}{dz})^2}. \tag{2.60}$$

If $Ri < 0$, then the flow is convectively unstable. If $0 < Ri < 1/4$ the flow is shear unstable.

2.6.1.3 Spontaneous imbalance

The spontaneous imbalance mechanism consists of emission of imbalanced motions (gravity waves) by a purely balanced flow (e.g., in geostrophic balance) without any external forcing mechanism. These two motions have a timescale separation that can be estimated by the Rossby number Ro. For large-scale flows, $Ro << 1$, meaning that these flows characterise most of the dynamics of the atmosphere and that the coupling with IGWs is weak. Although many balanced models, relying on this assumption, completely filter out IGWs, the interaction has important consequences, and the spontaneous emission is among them. The inevitability of this emission, although counterintuitive, has been longly discussed in the literature and finally proven by several authors. In the

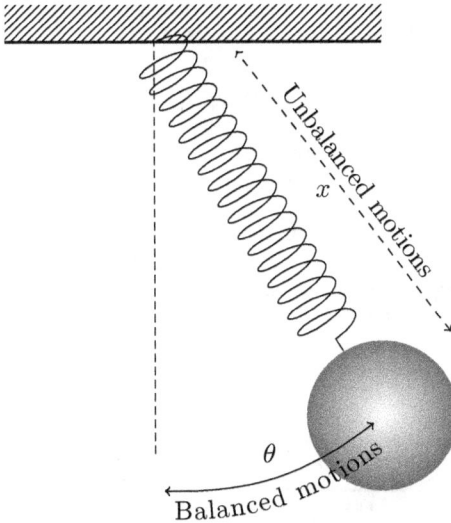

Figure 2.12: Drawn of the elastic pendulum after Vanneste (2013). The elastic pendulum serves as analogs for the slow balanced motion and fast IGWs of geophysical fluids. The two-timescale dynamical system is represented by the large natural frequency of vibration of the spring (with the slow angle θ) and the natural small frequency of oscillation of the pendulum (with fast extension x).

following we will explore the theoretical concepts of spontaneous generation mechanisms for two flow regimes: the $Ro \ll 1$ rapidly rotating regime, and the shallow-water stratified regime see Vanneste (2013) for a complete review).

$Ro \ll 1$ **regime.** Spontaneous emission in this regime is based on low-order models such as the Lorenz-Krishnamurty model describing with five ordinary differential equations the interactions of three slow vortical modes and two fast gravity wave modes. The equations are very similar to the ones of a system of a spring tied to a pendulum (see sketch in figure 2.12) that can, therefore, be used as an analog where the ratio of the slow pendulum (representing the slow balanced motion) to the fast spring oscillation (similar to IGWs) frequencies are equivalent to the Rossby number. This is a two-timescale systems, in which the interactions between the slow and fast degrees of freedom are weak. With this model, Vanneste and Yavneh (2004) have proven that the spontaneous generation of fast motions from slow balanced ones is inevitable, although the amplitude of these generated waves is exponentially small in Rossby number, i.e., of a form involving $\exp(-\alpha/Ro)$, with a prefactor that involves algebraic powers of the Rossby number Ro.

From a more practical point of view, in a flow where $Ro \ll 1$, regions in which $Ro > 1$ locally are most favourable for IGWs generation.

Rotating shallow-water $Ro > 1$ regime. Although so far spontaneous emission in stratified fluids could not be reduced to a single process, it is frequently connected to spontaneous emission in rotating shallow-water (RSW) flow. For RSW it can be related to Lighthill radiation, where the source is considered to be small compared to the emitted long waves and it is assumed that the Rossby number is $Ro > 1$ and the Froude number $Fr = U/\sqrt{gH} \ll 1$. It is interesting to know that this theory is an extension of Lighthill (1952) study of aerodynamically generated acoustic waves by turbulent vortical motions done by Ford (1994), who included rotation and therefore derived the inertia-gravity wave radiation term. An interesting analogy between the two is that as IGWs are filtered out in the quasi-geostrophic, sound waves are filtered out in incompressible fluids, when the Boussinesq approximation is made. By rearranging the equations of motions in the following way

$$\underbrace{\left(\frac{\partial^2}{\partial t^2} + f^2 - gH_0\nabla^2\right)\frac{\partial h}{\partial t}}_{\text{Linear operator}} = \underbrace{\frac{\partial}{\partial t}\nabla \cdot \vec{G} + f\vec{k} \cdot \nabla \times \vec{G} + \frac{g}{2}\frac{\partial}{\partial t}\nabla^2 h^2}_{\text{non-linear terms=Lighthill-Ford radiation}}, \tag{2.61}$$

where

$$\vec{G} = \vec{u}\nabla \cdot (h\vec{u}) + (h\vec{u} \cdot \nabla)\vec{u}, \tag{2.62}$$

with h the height of the free surface of the fluid, H_0 the height of the fluid at rest, and $\vec{k}$ the unit vertical vector (Williams et al. 2005). The term on the left-hand side of (2.61) consists of linear wave operator and is the gravity wave equation for fluid at rest forced by nonlinear terms on the right-hand side, in the form of a quadrupole. For emission to happen in a rotating shallow water system, the condition is that the Rossby number $Ro \geq 1$ and the Froude number $F = U/(gH)^{1/2} \ll 1$. The first condition is necessary since rotation inhibits the emission of waves and frequency matching between the vortical motion and the inertia-gravity waves only occurs for $Ro \geq 1$. The second condition allows asymptotic investigation of the problem and implies that there is a scale separation between the small balanced motion and the large-scale gravity waves that are emitted. Indeed, when $F \ll 1$, the non-linear right-hand side of (2.61) consists mainly of the non-divergent vortical flow, and it can, therefore, be regarded as a given source of inertia-gravity waves. This implies that unsteady vortical flows can emit freely propagating inertia-gravity waves whenever the Lighthill-Ford radiation term is different from zero.

Note that the wave source, considered to be small compared to the long gravity wave in shallow water, is different for the stratified case where IGWs are excited by slow balanced motion even if the condition $Ro > 1$ and $Fr \ll 1$ does not hold. Hence, caution should be exercised when comparing spontaneous emission in stratified fluids with Lighthill radiation.

2.6.1.4   Shear instability

Another nonorographic generation mechanism that is still poorly understood and even less clear than spontaneous emission is the excitation of gravity waves by unstable shears above the tropopause jets (Bühler et al. 1999). Because gravity waves generated by shear instability are located in the same atmospheric regions as the ones generated by spontaneous emission, it is not clear yet which mechanism is the dominant one (note that some authors consider shear instability to be a spontaneous emission process for non-rotating flows (Plougonven and Zhang 2014b)).

Moreover, this type of instability and the propagated gravity waves have been attributed to the observed small-scale waves in shear-driven fluid in rotating annulus experiments (Flór et al. 2011, Williams et al. 2008).

By quantifying the ratio between the buoyancy frequency and the vertical shear of the velocity, which is the definition of the Richardson number (2.60), it can be checked whether the fluid supports the shear instability or not. The Richardson number criterium divides the stable and the unstable regimes. A shear flow with $N^2 > 0$ is unstable if

$$0 < Ri < \frac{1}{4}. \tag{2.63}$$

2.6.2   Gravity waves propagation-wave capture

The exit region of the atmospheric jet is not only important for generation mechanisms, as we have previously seen, but plays also an important role for wave propagation. The propagation of gravity waves, regardless of the generation mechanism responsible for their emission, can strongly influence several characteristics of the waves themselves. In particular, gravity wave packets propagating through regions with strong horizontal wind deformationcan become trapped due to the so-called wave-capture mechanism (Plougonven and Snyder 2005), which we will briefly explain in the following.

To see the effects of wave capture on the wave properties we can look at a simplified theoretical model describing the linear propagation of a gravity wave packet in a background flow made of a superposition of horizontal deformation and vertical shear (Bühler and McIntyre 2005).

In particular, we use the ray equations (see derivation in appendix A.3)

$$\frac{d_g \vec{k}}{dt} = -\nabla \Omega \tag{2.64a}$$

$$\frac{d_g \omega}{dt} = \frac{\partial \Omega}{\partial t} \tag{2.64b}$$

$$\frac{d_g \vec{x}}{dt} = \vec{c}_g \tag{2.64c}$$

and consider a wave packet in a background flow $\vec{U}$. The doppler shifted dispersion relation for almost plane gravity waves is

$$\omega(\vec{x}, t) = \vec{U}(\vec{x}, t) \cdot \vec{k}(\vec{x}, t) \pm N(\vec{x}, t) \left(\frac{K_h(\vec{x}, t)}{K(\vec{x}, t)} \right) = \Omega(\vec{k}(\vec{x}, t), \vec{x}, t) \tag{2.65}$$

where the first term introduces in the equation the frequency shift due to the Doppler effect.

Let us assume that $N = \text{const}$, and we consider a flow aligned with the jet such as $\vec{U} = (U(x), 0, W(z))$ and $U_x = -W_z = a$ (following from the flow incompressibility). Under these conditions (2.64a), we have

$$\frac{d_g \vec{k}}{dt} = -(\nabla \vec{U}) \cdot \vec{k} = - \begin{bmatrix} ak & 0 & 0 \\ 0 & 0 & 0 \\ 0 & 0 & -am \end{bmatrix}. \tag{2.66}$$

The solutions to (2.66) are of the form

$$\vec{k}(t) = (k_0 \exp(-at), l_0, m_0 \exp(at)) \tag{2.67}$$

When the flow is horizontally compressing ($a < 0$), we can see that the wave vector $k \to \infty$ and the vertical wavenumber $m \to 0$. This implies that the wave vector $\vec{k}$ will align with the jet and the wave crests are perpendicular to it. Furthermore, the phase velocity of the captured waves is

$$c_p = \vec{U} + \frac{\omega}{K} \frac{\vec{k}}{K} = \begin{bmatrix} U + \frac{\omega}{K^2} k \\ V + \frac{\omega}{K^2} l \\ W + \frac{\omega}{K^2} m \end{bmatrix} \tag{2.68}$$

and the group velocity

$$c_g = \begin{bmatrix} U + N \frac{km^2}{K^3 K_H} \\ V + N \frac{lm^2}{K^3 K_H} \\ W + N \frac{K_H m}{K^3} \end{bmatrix} \tag{2.69}$$

which shows that, for $k \to \infty$ and $m \to \infty$ both propagate with the jet speed since ω is constant along the ray for $\Omega(\vec{k}(\vec{x}, t), \vec{x})$

Chapter 3

Laboratory experiment set-ups

"No one believes the simulation results except the one who performed the calculation, and everyone believes the experimental results except the one who performed the experiment."

− Patrick J. Roache −

3.1 Differentially heated rotating annulus - a bit of history

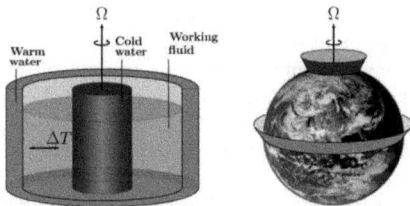

Figure 3.1: Laboratory experiment-Earth's atmosphere analogy

The first know laboratory experiment on the general atmospheric circulation dates back to the mid of the nineteenth century when Vettin (1857) conceived the idea of the differentially heated rotating annulus as a laboratory analogue to the planetary atmospheric system. Vettin's 'dishpan' experiment consisted of a rotating cylinder with ice placed in the centre to create a radial temperature contrast. Using air as working fluid and visualising the flow pattern with smoke from a cigar, Vettin firstly observed phenomena such as convective vortices and larger scale overturning circulation, obtaining the axisymmetric, now known as 'Hadley-type', circulation (Read et al. 2014). Unfortunately, his attempt to link the experimental observations to an understanding of meteorological phenomena was strongly criticised by prominent meteorologists who did not recognise the relevance of these experiments. In the same years, many other scientists showed an increased interest in the study of meteorological phenomena and started to develop theoretical models. Thomson (1892)(apparently unaware of the work of Vettin) proposed a very similar experiment using water as working fluid instead of air to demonstrate his theoretical model on the general circulation of the atmosphere. There is no trace that this experiment was ever built or further experimental attempts had been made until 1923 when Exner (1923) carried out experiments almost identical to the Vettin's ones and observed a system of vortices. For the experiment to be fully recognised as a valid model of the atmosphere, we have to wait until the extensive works by Fultz (1951) who used an open cylinder (similar to the dishpan) and Hide (1958) who, instead, used an annular experimental set-up. The latter apparatus is the progenitor of the classical differentially heated rotating annulus experiment, still used nowadays for studying the baroclinic instability mechanism which is at the base of heat and momentum transport at mid-latitudes. The apparatus and his analogy with the Earth's atmosphere are sketched in figure 3.1: an annulus with three concentric cylinders is mounted on a turntable so it rotates about its vertical axis of symmetry. The outer and inner rings, representing the equators and one of the poles respectively, are filled with hot and cold water and maintained at constant temperature. The thermally induced flow is studied in the water contained in the middle gap.

Over the years, experiments with this very simple configuration have lead to the understanding of the mechanisms driving the atmospheric circulation— i.e. gravity, rotation, and differential heating—and the separation from other mechanisms—i.e. the curvature of the Earth, the presence of water vapour in the atmosphere and the distribution of land and sea— that play a modifying but not essential role.

3.1.1   From the atmosphere to the lab - dynamical similarity

Since the atmospheric phenomena have extremely different scales form the equivalent laboratory experiments, the fundamental pre-requisite for the experiments to provide insights on such phenomena is to establish dynamical similarity. This means that, for the two flow systems to be comparable, they should have the same primary balance of acting forces, and, ideally, there should also be some similarity of boundary conditions. In general, it is very difficult if not impossible to reach a perfect similarity able to produce the identity of the field equations and boundary conditions by simply rescaling the parameters. However, the purpose of laboratory experiments is not to perfectly replicate the atmospheric system on a smaller size, but rather to get fundamental insights into the essential balance of forces and transport processes within the flow. Therefore, in most of the applications, many secondary terms can be neglected, so that the experiments are a somewhat simplified version of the atmosphere.

Much of the similarity problem is connected with that of the dimensional analysis in which the leading order balance is determined. By putting the field equations or boundary conditions in some suitable non-dimensional form, dimensionless parameters that define the ratio between acting forces emerge, as we have already discussed in section 2.3.1. We now want to rewrite the non-dimensional numbers that are relevant for the experiment in terms of the parameters of the annulus.

We start considering the **Rossby number** $Ro_T = U_T/(fL)$, where U_T is the typical velocity scale, $f = 2\Omega$ is the Coriolis parameter, and $L = (b-a)$ is the horizontal dimension in the annular geometry, i.e. the gap width. For the flow in the rotating annulus, the velocity is given by the thermal wind equation $U_T = (Hg\alpha\Delta T)/(fL)$, where ΔT is the externally imposed radial temperature difference, g is the acceleration of gravity, H the total fluid depth, and α the volumetric thermal expansion coefficient. By substituting U_T into the definition of Ro_T, we have

$$Ro_T = \frac{Hg\alpha\Delta T}{(2\Omega)^2(b-a)^2}.$$
(3.1)

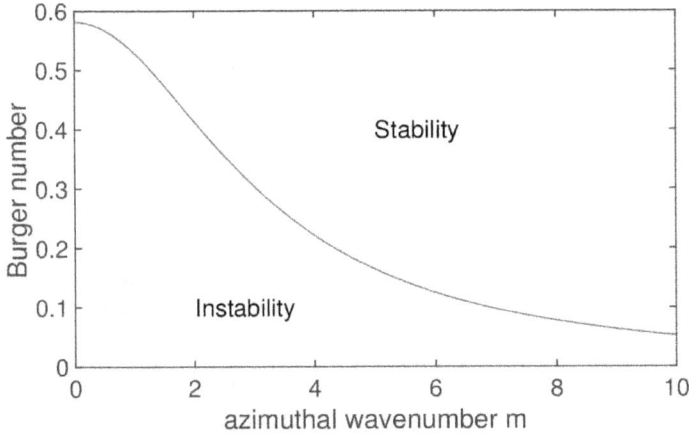

Figure 3.2: Dependency of the Burger number on the azimuthal wavenumber m (the radial wavenumber is kept constant $l = 1$).

Another significant parameter for the rotating annulus experiment—defining the ratio between Coriolis forces and viscous dissipation—is the **Taylor number**

$$Ta = \frac{4\Omega^2 (b - a)^5}{\nu^2 H},$$ (3.2)

where ν is the kinematic viscosity. These two numbers, together with the inverse aspect ratio of the annulus $\Pi = L/H$ and the Prandlt number $Pr = \nu/\alpha$ (ratio of the viscosity and thermal diffusivity), are fundamental to determine when the transition from axisymmetric flow to baroclinic instability can be expected. In section 2.4.1, we have seen that the condition for baroclinic instability, according to the Eady model, is $\mu < \mu_{\text{crit}} = 2.399$. Now we want to apply the model to the annular geometry and find the normal modes.

By using the definition of the Rossby radius of deformation L_D and using the **Burger number** (2.20), the instability criterium can be written as

$$Bu = \left(\frac{N}{f} \frac{H}{(b - a)} \right)^2 = \left(\frac{L_d}{(b - a)} \right)^2 < \left(\frac{\mu_D}{\pi} \right)^2,$$ (3.3)

where $(\mu_D/\pi)^2 = 0.583$ (Hide 1969).

Note that the critical Burger number, for which the instability occurs, decreases for higher wavenumbers (Drazin 1978)

$$Bu_c = \frac{0.583}{\left(l^2 + \frac{m^2}{\pi^2} \right)}$$ (3.4)

as it can be seen from the plot in figure 3.2.

We have seen in chapter 2 that viscosity can be neglected in the fluid interior and it becomes important only in thin layers at the walls: the boundary layers. In the thermally driven rotating annulus three different boundary layers types can be found. The first two are related to viscous effects and are: a horizontal boundary layer at the bottom, called 'Ekman boundary layer', and the vertical boundary layer situated at the inner and outer walls, called 'Stewartson boundary layer'. The thickness of these boundary layers is proportional to the Ekman number (2.14) and is

$$\delta_E = HEk^{\frac{1}{2}},$$ (3.5)

for the Ekman layer and

$$\delta_S = (b-a)Ek^{\frac{1}{3}},$$ (3.6)

for the viscous Stewartson layer at the side walls. The last boundary layer is related to the heat exchange between the lateral walls and the fluid. Its thickness is

$$\delta_T = H \left(\frac{\nu\kappa}{g\alpha\Delta TH^3} \right)^{\frac{1}{4}}.$$ (3.7)

3.1.2   Flow regime in the rotating annulus

Varying either the magnitude of the impressed temperature at the boundary or the rotation rate, four flow regimes (schematically listed in figure 3.3) can develop in the annular gap: axisymmetric flow, steady waves, vacillation and irregular flow. The only flow symmetric about the rotation axis is a steady zonal flow with a vertical shear, which is in thermal wind balance with the radial temperature gradient (see section 2.3.1). This axisymmetric flow can be found outside the anvil-shaped region (see figure 3.4) and is subdivided into 'upper axisymmetric regime' located in the region above for which the stratification suppresses the instability, and 'lower axisymmetric regime', located in the region below for which diffusive effects stabilise the zonal flow (Früh and Read 1997). This last regime can be reached only for low ΔT and high Ω.

Inside the anvil-shaped region, different forms of non-axisymmetric flow can occur; steady and vacillating waves are manifestations of baroclinic instability. Vacillating flow can consist in periodic changes in the shape of the wave ('wave-form vacillation'), in the wave amplitude ('amplitude vacillation'), or periodic alternations in azimuthal wavenumber ('wavenumber vacillation'). The azimuthal wavenumber m observable in the vacillating and steady wave regimes is restricted by the geometry of the system and more in particular by the dimensions of the gap. The empirical law—found by Hide and Mason (1970) expressing the minimum and maximum wavenumber $m_{min} \leq m \leq m_{max}$ that must be integers—is

$$\frac{1}{4}\pi\frac{(b+a)}{(b-a)} \leq m \leq \frac{3}{4}\pi\frac{(b+a)}{(b-a)}.$$ (3.8)

The flow regime can be determined by the dimensionless Taylor and thermal Rossby numbers. However, the wavenumber m is only statistically determined by Ta and Ro_T, since the flow exhibits hysteresis effects together with the existence of multiple equilibria that results in the observation of different wavenumbers or flow type under identical parameter conditions (Früh and Read 1997). The alternation of cyclonic and anticyclonic eddies is visible in the waves regime (see figure 3.3), and they move prograde with a well-defined drift speed.

For high rotation rate and high temperature difference the wavy flow transitions to a highly irregular and aperiodic flow regime.

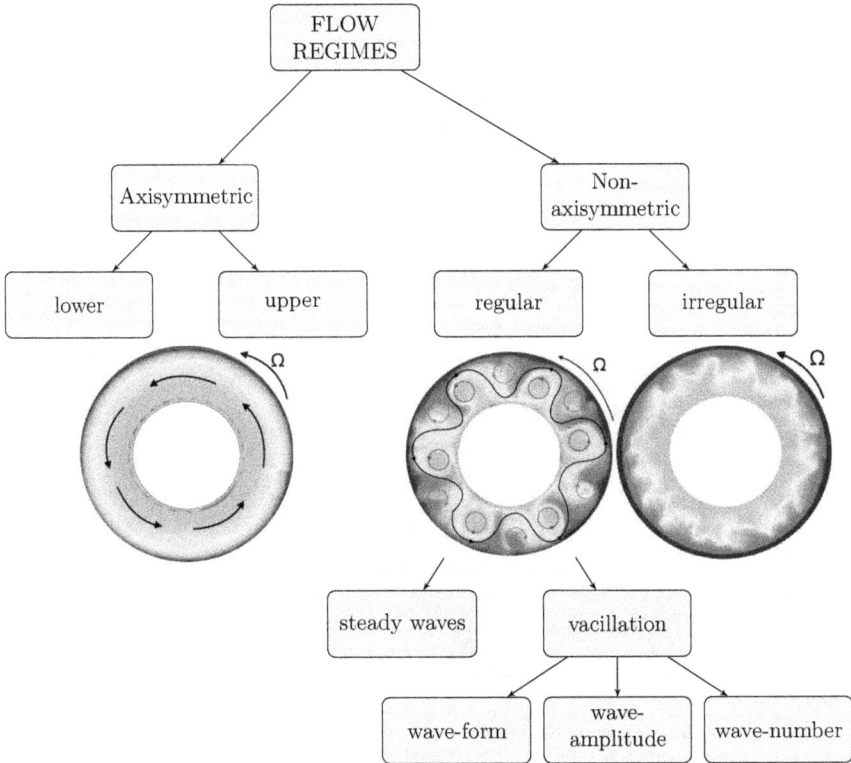

Figure 3.3: Diagram of the different free types of flow developing in a rotating fluid subject to axisymmetric differentially heating and cooling (drawn after Hide (1969)). The occurrence of the different flow types depending upon Ro and Ta is shown in the diagram regime in figure 3.4.

The experimentally measured transition points in the $Ta - Ro_T$ parameter space follow an anvil-shaped curve (see figure 3.4 reproducing a typical diagram regime, e.g., Hide and Fowlis (1965) and Hide and Mason (1975)). The inviscid Eady model, cannot reproduce the shape of the experimental diagram regime, giving instead a straight line corresponding to a constant value of Ro_T (red line plotted in figure 3.4). Nevertheless, the Eady model predicts fairly well the transition between the upper axisymmetric regime and the wave regime for experimental configuration where the flow is not dominated by viscous effects, as we will show in section 3.2. In order to reproduce the experimental curves, several theoretical works have extended the solutions of the Eady problem taking into account the effects of viscosity. Barcilon (1964) assumed that the effects of friction at the flat rigid bottom and top lid dominate the effects at the side walls. With this approximation, valid for shallow water fluids, dissipation is introduced only at the Ekman boundary layers at the top and the bottom, and in the fluid bulk the model converges to the Eady solution.

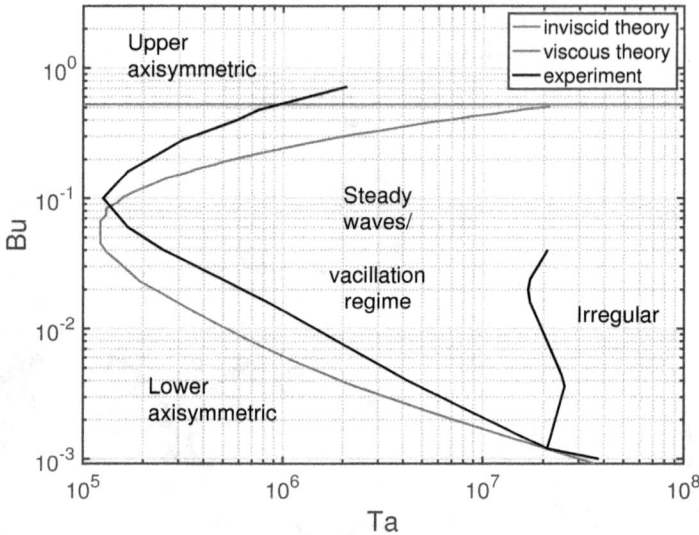

Figure 3.4: Sketch of the typical diagram regime for flow regimes in the thermally-driven baroclinic annulus (the black curve shows the experimental data measured by Hide and Mason (1970)). The three vacillation regimes listed in figure 3.3 are located inside the curve, together with the steady waves. The red straight line shows the inviscid instability criterium from Eady and the blue curve is the instability viscous criterium from O'Neil (1969).

Successively, O'Neil (1969) extended Barcilon (1964) theory by using a systematic analysis to include the effects of sidewalls on the instability results and to allow for a free surface. Her solution for the neutrally stable modes (equation 6.2 in O'Neil (1969)) is written as

$$\frac{m^2\sqrt{2}\sigma Q^2 Z(Q)}{(\sigma-1)(m^2+\pi^2)^2}\left[\Pi^2(m^2+\pi^2)\left(F(Q)+\frac{1}{\sigma-1}\right)+Q^2(1-F(Q))\right]^{-1}=-\frac{\epsilon^{3/2}}{\epsilon_B^2}. \quad (3.9)$$

$F(Q)$ is the interior solution

$$F(Q) =\frac{1}{4\cosh^2 Q}\sum_{\pm}\left\{e^{\pm 2(Q-\tanh Q)}\Re\left[E^{\mp}(2\tanh Q)-E^{\mp}(-2(Q-\tanh Q))\right]\right.$$
$$\left.+\log\left(\frac{Q-\tanh Q}{\tanh Q}\right)\right\}, \quad (3.10)$$

where $E^{\pm}$ are the exponential integral functions (Abramowitz and Stegun 1965) expressed as

$$E^{-}(x) = \int_{-\infty}^{x}\frac{e^{x'}dx'}{x'} = \begin{cases}Ei(x), & \text{if } x<0, \\ \overline{E}i(x)+i\pi, & \text{if } x>0.\end{cases} \quad (3.11)$$

and

$$E^+(x) = \int_\infty^x \frac{e^{-x'} dx'}{x'} + i\pi = \begin{cases} \overline{E}i(-x), & \text{if } x < 0, \\ Ei(-x) + i\pi, & \text{if } x > 0. \end{cases} \tag{3.12}$$

The function $Z(Q)$ is related to the boundary condition for a free upper surface

$$\gamma^2 Z(Q) = -\frac{\epsilon^{3/2}Q^2}{\sqrt{2}} \left(1 - \frac{1}{\sigma}\right) \left[\Pi(m^2 + \pi^2) \left(F(Q) + \frac{1}{\sigma - 1} \right) + Q^2(1 - F(Q)) \right] \tag{3.13}$$

where $\gamma = m\epsilon_B Bu$.

The left-hand side of (3.9) is a function of the geometry of the annulus, the Prandlt number $\sigma = \nu/\kappa$, the wavenumber m, and the variable Q that is related to the Burger number $Q = \sqrt{Bu(m^2 + \pi^2)}$. The right-hand side has two variables: $\epsilon_B = \Delta T/\Delta T_v$ and the Ekman number $\epsilon = \nu/(2\Omega h^2)$, which is related to the Taylor number since $\epsilon = \Pi^{-5/2}(Ta)^{-1/2}$. Once the values of the parameters Π, m, and ϵ_B are fixed, the equation can be solved using the shooting method (Press et al. 2007) and this gives the neutral stability curves. The solutions are valid when the inverse aspect ratio $\Pi < 1$, which implies a wide annular gap and shallow water.

Not many attempts have been made to run experiments with $\Pi < 1$. In most cases, the differentially heated rotating annulus is a table size experiment, and the gap has a width varying from $\mathcal{O}(10\,\text{cm})$ to $\mathcal{O}(1\,\text{cm})$. This is the case of the classic experiment carried out by Hide and Mason (1970), which has $\Pi > 1$ and, therefore, the model proposed by O'Neil (1969) is not valid for this set-up strictly speaking. Nevertheless, the plot of the theoretical curves for the Hide's experimental parameters in figure 3.4 shows that the shape of the regime diagram is very well captured and the model's values are not too far from the experiment, even in this case. We will compare the transition predicted by the O'Neil's model for experiments having $\Pi < 1$ in section 3.2 when discussing the results obtained with the MSGWs tank.

Before moving to the description of the experimental apparatuses used for this thesis work, it is worth mentioning for completeness that another class of experiments (introduced by Hart (1972)) can be used to investigate the baroclinic instability. The apparatus, sketched in figure 3.5(a) and compared to the differentially heated rotating annulus set-up (b), consists of a rotating cylinder filled with two immiscible fluids having different densities, where the baroclinic instability is driven mechanically by a differently rotating lid. Both systems are forced by the vertical velocity shear, which is driven by the thermal wind balance in the first configuration and by the rotating lid in the second one. In the latter experiment, the baroclinicity is concentrated at the interface between the two fluids. The experiment is an analogue of linear and weakly non-linear instabilities in the two-layer rotating flow systems theoretical work of Phillips (1954) and Pedlosky (1970).

The mechanically driven system has the advantage of having a more straightforward forcing mechanism and that it does not involve complex boundary layer circulations. Therefore, theoretical models are easier to compare and verify experimentally via this set-up (Read et al. 2014). However, for this experiment the analogy with the atmosphere is not as clear as the physical phenomena are different. In recent years, the two-layers annulus has been used by Lovegrove et al. (2000), Williams et al. (2005), and Williams et al. (2008) to study the emission of interfacial gravity-waves from baroclinic waves, similarly to what we aim to do in this thesis. A qualitative comparison between the

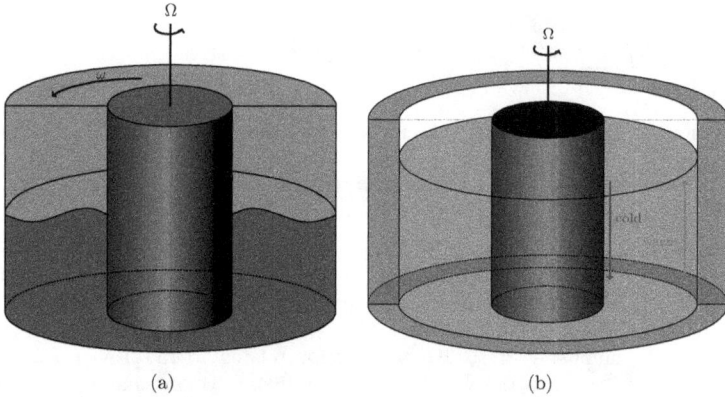

(a)   (b)

Figure 3.5: Two laboratory experiments to generate baroclinic waves: mechanically driven with two-layer stratification (a) and thermally driven with continuous stratification (b).

gravity waves occurring in the two types of experiments will be given in section 7.2.1, where the gravity wave properties observed in our experiment are discussed.

3.2   Rotating annulus: two experimental apparatuses to study IGWs

In this section of the thesis, the two differentially heated rotating annuli experiments, operating at the BTU laboratories, are described. Before giving the details of this two experimental set-ups, a short explanation of why the classical rotating annulus configuration is not particularly favourable to investigate emission of gravity waves from baroclinic jets is given together with the motivation for the need of the two modified experiments. This section is part of the paper 'A new atmospheric-like differentially heated rotating annulus configuration to study gravity wave emission from jets and fronts.' submitted to the journal 'Experiments in Fluids'.

The classical configuration of the differentially heated rotating annulus experiment used in laboratories has, in meteorological terms, an unusual inverse aspect ratio ($\Pi = H/L$, the ratio between the vertical depth and the horizontal scale of motion) greater than one. The reason why this ratio is considered unusual is that its typical values in the atmosphere are much less than one. One reason why experiments with $\Pi < 1$ have rarely been performed is that, when the total fluid depth is decreased, the wave regime becomes less accessible due to the increase in Ekman dissipation (Miller and Fowlis 1985). Although this geometrical difference between the experiment and the atmosphere is not primarily significant for the study of large-scale waves, it becomes fundamental when investigating IGWs. Indeed, the narrow gap classical configuration leads to a ratio between the buoyancy frequency N and the Coriolis frequency f less than one. Because this ratio is much greater than one in the atmosphere, it can be expected that IGWs properties—which depend on N and f according to their dispersion relation (2.51)—differ between experiment and atmosphere. For $N/f < 1$ waves can still propagate in the flow, but they are inertial waves, which are rotation dominated, and differ from gravity waves in several aspects. Since our aim is to study atmospheric jet generated waves, which are gravity dominated, the only way to obtain similar features is to use a configuration of the experiment that allows $N/f > 1$.

Hide and Fowlis (1965) and Lambert and Snyder (1966) investigated the effects of changing the aspect ratio (for $\Pi > 2$) on the flow regime in a differentially heated rotating annulus. They found that the stability curve, which determines when baroclinic instability occurs, is affected by the parameter Π. By going into more detail, they show that the horizontal viscous boundary layers are essential in determining the stability of the curve. Douglas and Mason (1973) extended the investigation to $\Pi < 1$ by using an apparatus with large gap width, which allowed for the study for the study of a system with small inverse aspect ratio. Differently from the previously mentioned studies, their experiment was covered by a lid and hence had a rigid bottom and top surface, as it is also assumed in analytic theories of baroclinic instability formulated by Eady (1949) and Barcilon (1964). The comparison of critical values from the experiment with linear theory showed a reasonable agreement. What emerges from these studies is that, in the classical configuration with the tall annuli, the Ekman layer is small compared to the total fluid depth and the large-scale flow is not affected by Ekman friction. When the fluid depth is decreased on the other hand, friction becomes more and more dominant and can, therefore, not longer be neglected.

The purpose of this section is to demonstrate that a baroclinic wave experiment with $N/f > 1$ is possible only with a small Π. However, to avoid ending up in an Ekman dominated flow, H must be at least in the order of a few centimetres. This, in turn, implicates an L in the order of decimetres. Hence, studies on gravity wave emission from baroclinic waves in an atmosphere-like regime with $N/f > 1$ can only be done in an experiment that differs in size and aspect ratio from the classical baroclinic wave experiments (Hide and Fowlis 1965).

We now make the assumption that $\Delta_z T \approx \Delta T$, where $\Delta_r T = (b-a)\partial T/\partial r$ is the radial temperature gradient in the fluid interior and $\Delta T = T_b - T_a$ is the imposed lateral temperature difference (we shall discuss the validity of this assumption in section 7.1.1). For this assumption, the Burger (3.3) and the thermal Rossby number (3.1) are equivalent. Another consequence of making this assumption is that the buoyancy frequency, which depends on the stratification of the fluid, can be written as $N^2 = g\alpha\Delta T/H$. In this way, the Burger number can be rewritten as

$$Bu = \frac{Hg\alpha\Delta T}{(2\Omega)^2(b-a)^2} = \left(\frac{N}{f}\right)^2 \left(\frac{H}{L}\right)^2. \tag{3.14}$$

Shallower systems have a greater value of N for baroclinic waves, given the same imposed radial temperature difference. It can, therefore, be seen how the geometry of the rotating annulus becomes crucial when using it for the study of internal gravity waves, the frequencies of which strongly depend upon N. A similar argumentation about the importance of geometry in numerical simulations has been discussed by (Borchert et al. 2014).

One of the generation mechanisms of internal gravity waves that is still not fully understood is spontaneous emission (see section 2.6.1.3). The differentially heated rotating annulus represents a suitable laboratory experiment to investigate spontaneous emission of internal gravity waves from baroclinic waves and jets. Although, spontaneous emission in stratified fluids could not be reduced to a single process thus yet, it is however frequently connected to spontaneous emission in rotating shallow-water (RSW) flow. For RSW it can be related to Lighthill radiation, where the source is considered to be small compared to the emitted long waves and it is assumed that the Rossby number is $Ro > 1$ and the Froude number $Fr = U/(gH)^{1/2} << 1$ (Vanneste 2013, Sugimoto et al. 2008). Lighthill radiation of gravity waves is based on the isomorphism between compressible waves and shallow water surface waves. Studying baroclinic waves in a continuously stratified fluid, as we do in the differentially heated rotating annulus, this isomorphism breaks down but spontaneous gravity wave emission is still possible, although it differs from Lighthill radiation. Since in stratified flows $Fr = U/(NH)$, and hence $Bu = (Ro/Fr)^2$, we find that in flow regimes with $\Delta_z T \approx \Delta T$ the Froude number depends on the thermal Rossby number via $Fr \approx Ro^{1/2}$. This implies that in stratified flows the background condition typical for the RSW Lighthill radiation cannot be reached if the vertical temperature difference equals the lateral temperature difference. Therefore, it is not yet clear at which region in the $Ro - Fr$-space we can expect the most effective gravity wave radiation.

Figure 3.6(a) shows, for fixed $Bu = 0.6$ (i.e. where baroclinic instability is predicted by the Eady model, according to (3.3)) and $Fr = 0.77$, the variation of N as a function of the rotation rate Ω, calculated with (3.14) for different experimental set-ups tested in the laboratories at the BTU. The label 'small-tank' refers to the classical set-up (see table 3.1 for more details) with maximum and minimum fluid depths. The label 'big-tank' refers, instead, to the new experiment (see table 3.2 and section 3.3) with characteristics more similar to the atmosphere. For comparison, the thermohaline configuration of the small-tank experiment is also shown in the plot and labelled as 'barostrat' (see section 3.2.1 and chapter 6). This experiment, for which $N/f > 1$ is reached by introducing a vertical salinity gradient, represents a special case since the baroclinically unstable layer lies on top of a stably stratified one. For this reason, it has no bottom Ekman layer influencing the flow. Rodda et al. (2018) observed inertia-gravity waves trapped along the baroclinic jet and showed that in a sense barostrat is, for some aspects, an ideal experiment to study emission and propagation because of the formation of alternating baroclinically unstable and stable layers qualitatively resembling the troposphere and stratosphere (the results are reported in chapter 6). The latter is too stably stratified for baroclinic instability.

As it can be seen in figure 3.6(a), the black line for the small tank with fluid depth $z = 13.5$ cm is below the line $N/f = 1$ (in purple) whereas the red line for the atmosphere-like big tank is above that curve. Hence, with the big tank we are closer to the atmospheric case since $f < \omega < N$. Note that we have about $N/f = 2$ which is still small compared to the atmosphere where $N/f = 100$. It is, moreover, relevant to note that for typical laboratory conditions, which give $N \approx 0.5$ rad/s, the big tank rotates much slower compared to the small tank. Further, note that for all lines the thermal Rossby number is fixed to $Ro_T = 0.60$ and that $Fr = Ro_T^{1/2}$. By reducing the water depth in the small tank to $z = 3$ cm, which is the shallowest configuration for which a regular baroclinic wave was observed

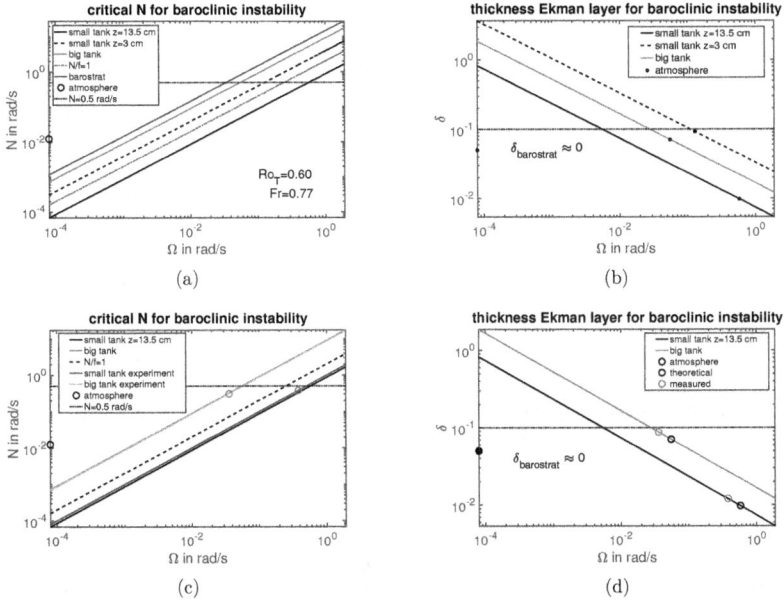

Figure 3.6: Critical buoyancy frequency N as a function of the rotation rate Ω calculated with (3.14) for different experiments and experimental configurations theoretical values(a) and experimental values (c). Ratio of the Ekman layer thickness and the fluid depth, δ, as a function of Ω theoretical values (b) and experimental values (d).

(see figure 3.8), N/f raises to 1.2; however this is, in fact, an unfortunate configuration for investigating IGWs since the frequency band is rather narrow. Finally, for the barostrat experiment, N/f in the baroclinically unstable layer is even greater than for the big tank. The question is how thin the fluid layer may be before Ekman effects influence baroclinic instability.

What we can already see from the qualitative analysis of the different experiments is that the greater the inverse aspect ratio Π, the more suitable the experimental configurations are to study IGWs. However, when the fluid depth is decreased in the laboratory experiments, viscous effects at the bottom and, in case of a rigid lid, top of the tank cannot be neglected. To estimate the importance of friction, we have plotted in figure 3.6(b) the ratio of the Ekman layer thickness ($\delta_E = (\nu/f)^{1/2}$ where ν is the coefficient of the kinematic viscosity) and the fluid depth $\delta = \delta_E/H$ as a function of Ω. The circles display values for $N \approx 0.5$ rad/s corresponding with the intersection of the horizontal dashed line with the sloping lines of the different experiments in figure 3.6(a). The black line for the small tank (with the fluid depth $d_{small} = 13.5$ cm, mostly used for the experiments) is far below the line $\delta = 1$. For the big tank experiment, we find $\delta \approx 0.1$ that corresponds roughly with the (turbulent) atmosphere. For $h = 3$ cm Ekman layers cover about 10% of the fluid depth. As we will see later, this seems to be the limit for the development of regular baroclinic waves close to the transition point. As mentioned earlier, barostrat has no Ekman layer (no rigid bottom boundary for the thin layer close to the surface) and is, therefore, an ideal set-up. It should be noted that even outside the boundary layers the flow in the annulus is 'more viscous' than the atmosphere, i.e. its characteristic Reynolds number is much smaller than the atmospheric one. This theoretical consideration can be validated by running some laboratory experiments in which the parameters for which the transition between axisymmetric and wave regime occurs are measured and from that the Burger number is calculated. We have done this for the big- and small-tank experiments and by fixing the experimentally obtained Burger number, we can plot again the critical buoyancy frequency N as a function of the rotation rate Ω (see figure 3.6(c)) and the ratio of the Ekman layer thickness and the total fluid depth as a function of Ω (see figure 3.6(d)). The proximity of the experimental lines to the theoretical lines marked in figure 3.6(c) confirms how the inviscid theory works fairly well in predicting the transition points for both experiments. In figure 3.6(d) it can be seen that the theory gives smaller values for δ. The measured ones show that for the big-tank, the ratio between the boundary layer and the total fluid depth is close to the critical value for which we expect significant interference with the flow.

We can conclude that (apart from barostrat which is technically more complicated), only the big tank system fulfil the requirement $N/f > 1$ and $\delta \leq 0.1$. Next, we study boundary layer effects on baroclinic instability experimentally to confirm the critical δ.

To experimentally show that the classical apparatus is not suitable for shallow water configurations, we used the small tank to investigate the observed transition from axisymmetric to wave regime for decreasing water depths and compared it with the one predicted by the theoretical model from O'Neil (1969).

The experiments were conducted in the following way: initially, the outer and inner annulus were set to the wished (warm and cold) temperature, and the experiment was let warm up for 1 hour until it reached a constant radial $\Delta_r T$. Then rotation was switched on and increased stepwise; at each step, we waited for 20 minutes so that the experiment could settle down at a certain baroclinically unstable state. The flow was monitored with an infrared camera which allowed to measure at which rotation rate the flow regime passed from 'axisymmetric' to 'regular wave'. The same experiment was repeated systematically changing the fluid depth from $d = 1$ to 8 cm.

The Burger number for the first observed wave is plotted with blue circles in figure 3.7 for the different inverse aspect ratio (II); the fitted black dashed curve is compared with the theoretical inviscid Eady model (straight blue line) and the O'Neil viscous model (red curve). Both theoretical models are plotted for baroclinic instability $m = 3$, which corresponds to the wavenumber observed in the laboratory experiment. It can be seen that the O'Neil curve converges to the Eady model for deeper fluid and deviates from this model when going to shallower water systems, as expected. The experimental points are all situated between the two models, except for the shallowest configuration explored. Interestingly, for the experiments that were run with water depths $z \geq 3$ cm, we observed a fully developed baroclinic wave, and their points in figure 3.7 are noticeably closer to the Eady straight line than the two shallowest runs. For these two, having fluid depths of $z = 1$ and 2 cm, a regular wave regime never sets in and the transition points are marked where we observed the flow to deviate from axisymmetric. For $z = 2$ cm a wavelike instability, confined close to the inner cylinder, was observed (see figure 3.8(b)), but it did not extend to the full gap width even after waiting for one hour. For $z = 1$ cm, the

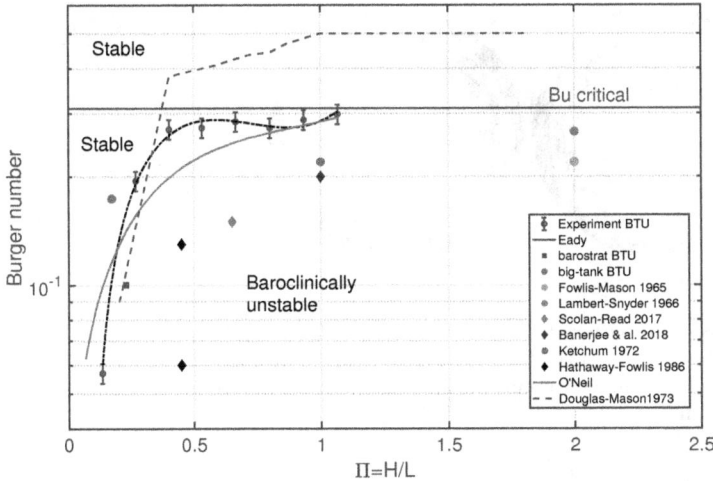

Figure 3.7: Plot of the transition from axisymmetric to wave regime for varying fluid depths. The straight blue line shows the (Eady model) inviscid criterium for $m = 3$ and $l = 1$. The red curve shows the transition according to O'Neil (1969) for $m = 3$ to be compared with the black dashed curve which is the fit of the experimental data (blue circles). The error bars show the maximum ΔT variation for different experiments. The region marked as 'stable' is to be intended as stable for $m = 3$ mode, but within it the front can still be unstable with respect to modes with $m < 3$. For fluid depths $d = 2$ cm and $d = 1$ cm the regime was not regular waves, rather geostrophic turbulence or more irregular waves (see figure 3.8). The blue dashed curve shows measured data for a rotating annulus with a rigid upper surface from Douglas and Mason (1973). The red and dark yellow dots are points taken from experiments from Lambert and Snyder (1966) and Hide and Fowlis (1965) respectively; note that in both cases the observed wavenumber is $m = 3$ (as for our experiment), but the aspect ratio of their experiment is $\Pi = 2$. The green dot corresponds to the transition observed with the atmosphere-like tank at the BTU. The purple dot corresponds to an experiment done by Ketchum (1972). The diamond-shaped markers are showing data from rotating annulus experiments with so-called horizontal convection temperature boundary condition (red (Scolan and Read 2017), black with $Bu = 0.13$ number $\Delta T_v = -1$, smaller $Bu = 0.06$ for $\Delta T_v = -1$ see Hathaway and Fowlis (1986), and blue (Banerjee et al. 2018)). The squared purple marker shows the transition for the upper layer in the barostrat experiment.

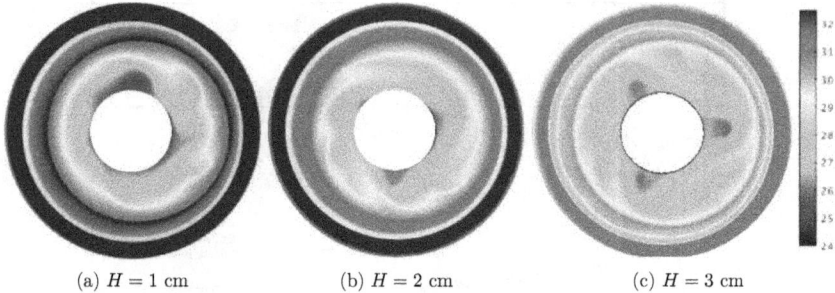

(a) $H = 1$ cm   (b) $H = 2$ cm   (c) $H = 3$ cm

Figure 3.8: Flow pattern for the three shallowest fluid depths visualised with the infrared camera. For $H > 3$ cm the flow regime is a regular wave $m = 3$ similar to the one in (c). The changes in temperature for the different experiments are taken into account with the error bars in figure 3.7.

first regime observed after the axisymmetric one already seems to be turbulent (see figure 3.8(c)). We increased the rotation rate until $\Omega = 7$ rpm, but could not observe a regular wave. Note that in our analysis we focused on $m = 3$ and we do not know the behaviour for other modes, for example $m = 1, 2$, which are in theory possible but not observed in the experiments we run. However, these modes exist in regimes where the rotation rate is lower than the one for $m = 3$ and hence where the Ekman layer is even thicker. Therefore, we expect a qualitatively similar behaviour as the one shown in figure 3.7 for $m = 1, 2$ and that the deviation from the Eady model starts for larger Π. Higher modes might survive for smaller Π.

Our data are further compared with other typical transition points for shallow-water differentially heated rotating annulus experiments present in the literature, plotted in figure 3.7. The dashed blue curve is taken from Douglas and Mason (1973) and represents the transition for an experimental set-up with a rigid upper boundary. In their work, Douglas and Mason (1973) studied the effect of varying the aspect ratio and compared experiments with the theoretical model by Barcilon (1964), which corresponds to their rigid-lid apparatus. They reported an increase in the azimuthal baroclinic wavenumber measured at the transition point with the decrease of the fluid depth set for the experiment. This might be explained by larger Taylor numbers and smaller Rossby deformation radii for smaller depth. The lid at the upper boundary affects the stability of the flow by shifting the transition of the baroclinic instability to higher Burger number values. Therefore, it has a destabilising effect in agreement with what was observed by Fein (1973) although the upper boundary condition does not substantially alter the baroclinic dynamics. Figure 3.8(a) shows the possible remains of such a large wave number case in an already turbulent state.

The second set of experiments included in figure 3.7 are performed in set-ups having a free surface and lateral differentially heating forcing system, performed by Hide and Fowlis (1965) (dark yellow dot in figure 3.7), Lambert and Snyder (1966) (red dot), and Ketchum (1972) (purple dot). We compare them with the transition for the atmosphere-like tank at the BTU (full details about the transition for this experiment will be given in section 7) marked by the green dot. The first two authors only investigated the region with $\Pi > 1$ for which the theory of O'Neil cannot be applied, whilst the last author reduced the aspect ratio to $\Pi = 1$. Note that Bu has been re-calculated according to (3.3), since some authors used different definitions.

The diamond shaped markers show three experimental set-ups for the rotating annulus with different variations of the so-called horizontal convection boundary conditions. The blue data is from Banerjee et al. (2018), for a rotating annulus with a peripheral spot heating system at the bottom on the outer edge and uniform cooling on the inner edge. Another experiment with peripheral local heating at the bottom, but with central cooling by a circular disk placed at the top was proposed by Scolan and Read (2017) and is indicated in figure 3.7 by the red diamond shaped marker. The last experiment of this kind is a configuration proposed by Hathaway and Fowlis (1986) (black markers) where the radial temperature gradients are imposed on a thermally conducting lid and bottom walls, and the side walls are insulating. The two points are taken from the horizontal difference of temperature $\Delta T_H = 10\,\mathrm{K}$. One with $\Delta T_v = -1$, which gives $Bu = 0.13$, and one with $\Delta T_v = -1$, which gives a smaller $Bu = 0.06$. It seems that in this case, the transition to baroclinic instability occurs for smaller Bu compared to the predictions by Eady and O'Neil for the case of differentially heating at the cylinder walls. These last two mentioned experiments have a similar horizontal scale, each more than one order of magnitude larger than the classical set-up.

Finally, the purple filled square marker indicates the barostrat experiment (Rodda et al. 2018). Although the value of Π for this point corresponds to the one for the 'classical' experiment where no fully developed baroclinic wave could be observed (see figure 3.8 (b) and (c)), for the barostrat experiment, a steady wave with $m = 3$ was found. This further confirms that the influence of the bottom Ekman layer inhibits the formation of baroclinic waves. Indeed, the shallow top layer of the barostrat experiment is sandwiched between the free surface and a stably stratified layer; therefore it has no destructive bottom Ekman layer.

In summary, from the results discussed so far we can evince that a tabletop-sized experiment, which is the usual configuration for the differentially heated rotating annulus in laboratories, is unfavourable to study atmosphere-like IGWs. Indeed, these waves require $N/f > 1$ which translate in a demand for $\Pi \ll 1$; as elucidated by figure 3.7 and 3.8 this will result in thin and viscous-dominated layers, which do not allow proper development of baroclinic waves in the classical experimental configuration.

We have experimentally proven how, for annuli having a gap width of the order of centimetres, it is impossible to reach a low value of the aspect ratio without having to deal with the viscous Ekman layer, which inhibits the formation of a steady baroclinic wave. By increasing the horizontal dimensions of the experiment, we obtained an aspect ratio equal to half the one of the classical configuration for a fluid depth $z = 6\,\mathrm{cm}$, which granted the formation of baroclinic waves undisturbed by the bottom viscous Ekman layer. Although we did not investigate with the big-tank in such a systematic way the influence of the total fluid thickness on the flow regimes, we did run some experiments with $H = 4\,\mathrm{cm}$ and for those a regular baroclinic flow was never observed. This confirms what we already observed that with $H = 6\,\mathrm{cm}$ we are almost at the critical value, and by further decreasing the fluid depth the boundary layer dynamics strongly influence the flow in the bulk.

Moreover, by decreasing the aspect ratio, the baroclinic instability regime is reached for higher values of N/f, which is a crucial point for the generation and propagation

Figure 3.9: Picture of the small differentially heated rotating annulus experiment at the BTU laboratories. This apparatus is used to run the barostrat experiments (see text for more details).

of IGWs. Indeed, this results in a broader band of intrinsic frequencies for the waves, meaning that we can expect to observe a broader spectrum of propagating and trapped waves in the experiment.

3.2.1 Barostrat

In this section we describe more in detail the thermohaline version of the differentially heated rotating annulus: the 'barostrat' experiment introduced by Vincze et al. (2016). A study of the different wave types and wave regimes developing in this configuration together with the possible large-scale small-scale wave interactions is reported in chapter 6.

The apparatus, visible in figure 3.9, is the classical annular tank which has been running at the BTU laboratories for the last fifteen years to investigate baroclinic wave properties (Von Larcher and Egbers 2005, Vincze et al. 2015)) and was also used by the author for the experiments reported in the previous section to investigate shallow water regime. The tank consists of three concentric cylinders and it is mounted on a turntable, which rotates around its vertical axis of symmetry. The middle annular cavity has a flat bottom topography whilst its surface is free. The cylindrical walls have a thickness

Geometrical and experimental parameters small annulus		
(a) Geometric dimensions		
Inner radius	a (mm)	45
Outer radius	b (mm)	120
Gap width	$b - a$ (mm)	75
Fluid depth	D (mm)	135
(b) Experimental conditions		
Difference of temperature range	ΔT (K)	$0 - 10$
Rotation rate range	Ω (rpm)	$0.1 - 5$
Rotation rate increment	$\Delta\Omega$ (rpm)	0.1
(c) Fluid's physical properties (de-ionised water)		
Density	ρ (kg m^{-3})	$1000 - 1170$
Kinetic viscosity	ν (m^{2}s^{-1})	1.004×10^{-6}
Thermal conductivity	κ (m^{2}s^{-1})	0.1434×10^{-6}
Exp. coefficient	α (K^{-1})	$0.207 - 0.327 \times 10^{-3}$
Prandtl number	Pr	7
Azimuthal wavenumber	m_{min}	2
Azimuthal wavenumber	m_{max}	5

Table 3.1: Parameters used for the barostrat laboratory experiment.

of 5 mm each; the inner wall is made of black anodised aluminium, whilst the middle and outer walls are made of borosilicate glass, so they are entirely transparent and fully accessible to PIV measurement (see section 4.2). The cold and warm baths are realised by circulating cold water pumped in the inner chamber and heating the fluid in the outer chamber by heating wires respectively. The heating and cooling system, as well as the rotation rate of the apparatus, are controlled by a self-developed experiment software programmed in LabVIEW®.

More details about the geometric parameters of the experimental setup, typical experiment conditions, and physical properties of the fluid are shown in table 3.1.

The difference between the classical and the barostrat experiment is that in the latter a layer of water initially stratified with salt instead of de-ionised water as working fluid is used. The density stratification is obtained before running the experiments with the double bucket method, introduced by Oster and Yamamoto (1963). The system used for our experiment at the BTU (see the sketch in figure 3.10) consists of two vessels having a volume of approximately 3 L each and containing the first fresh and the second saltwater. The two buckets sit on a platform at equal height, and a U-tube, running beneath the two vessels, joins them. A second tube, with smaller diameter, connects the freshwater bucket to the middle chamber of the experiment placed at a lower height. The procedure to obtain a linear stratified profile in the experiment is the following: first of all, the two buckets are filled with fresh water to the same height. The tubes are closed with stoppers to avoid mixing at this point. In the saltwater bucket, salt is added to achieve a desired density of the experiment parameter. For every gram of salt that was added to the saltwater bucket, 0.6 mL of fresh water is added to the fresh water bucket, so that the height of the water column in the freshwater bucket is somewhat higher than the height of the water column in the salt bucket. In this way, we avoid saltwater to rush into the fresh

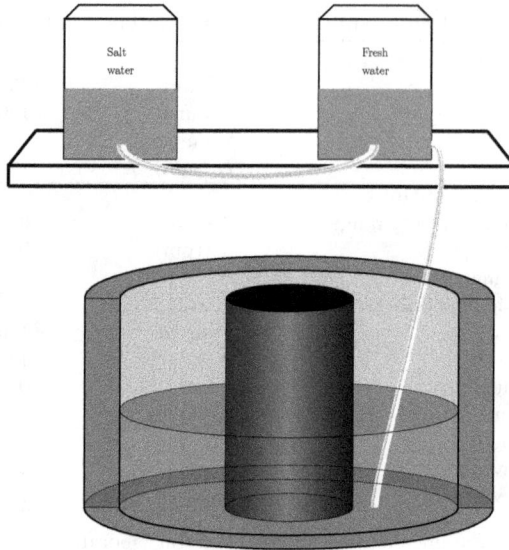

Figure 3.10: Sketch of the two bucket technique used to create the density profiles for the barostrat experiment. A bucket containing salt water (left-hand side) is connected to a bucket containing fresh water (right-hand side). The water with gradually increased density is then used to fill in the middle gap of the annulus.

water bucket when the stoppers are removed. To get a strong stratification, we added 100 g of salt every 100 g of water, which gives 33 g per 100 g. Since the solubility of NaCl in water at 20 °C (laboratory room temperature) is 36 g per 100 g water, the saltwater obtained was an almost saturated solution. The solution was then mixed with a magnetic stirrer for about one hour, until the salt was dissolved. At this point, the stoppers are removed so that saltwater inflows through the U-tube into the fresh water bucket yielding to a mixture of ever-increasing salinity (and therefore density) in time. In this way, the tank is filled through the pipe fixed at the bottom with progressively denser water.

The prepared stable vertical salinity profiles have been measured with a conductivity meter at the beginning of each experiment, before starting the rotation and at the end of it, after stopping the rotation. Successively, the conductivity has been converted to density using the empiric formula

$$
\begin{aligned}
\rho = {} & 9.22584 \exp(-11) \cdot C^4 - 2.94733 \exp(-8) \cdot C^3 \\
& + 4.19348 \exp(-6) \cdot C^2 + 0.000322098 \cdot C + 0.999075.
\end{aligned}
\tag{3.15}
$$

A sketch of experimental set-up with the double-diffusive convection developing in the barostrat experiment is shown in figure 3.11. Double diffusion is a mechanism that depends on two conditions, namely the buoyancy has to be a function of two variables (from which the term 'double') and these two variables diffuse at different rates (from which 'diffusion') (Smyth and Carpenter 2019). In our experiment, convection is observed in thin cells confined near the water surface and the bottom of the tank in regions where

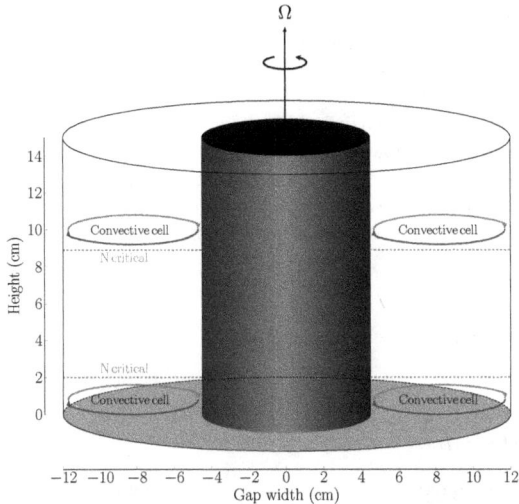

Figure 3.11: Sketch of the double-diffusive convection in the barostrat experiment. The red dashed lines located at the heights $z_{top} = 8.9$ cm and $z_{bottom} = 2$ cm indicate the interfaces between the convective regions (at the top and bottom of the tank) and the calm region at mid-depths

$N < N_{\text{critical}}$. For $N > N_{\text{critical}}$, convection is inhibited and no baroclinic instability can be observed.

An example of the typical vertical density profiles are plotted in figure 3.12(a), from which we can distinguish three layers in the tank: two shallow layers (about a few centimetres, indicated by the grey colour), on the top and on the bottom of the tank, both characterised by constant density, and a middle layer with an almost linear density profile. The sharp interfaces separating the regions, visible in the plot, are caused by convection. The plots for the density and the buoyancy frequency N (figure 3.12) are in excellent agreement with the ones shown by Vincze et al. (2016). The boundary conditions of no-flux for salinity at the surface and the bottom of the water column imposes a zero concentration gradient at these boundaries even before a temperature gradient is applied. In this configuration, only the two separate shallow fluid layers can be baroclinically destabilised. Indeed, the vertical salt stratification opposes the thermal convective motions until the ratio of the (horizontal) thermal density difference and the (vertical) salinity-induced density difference exceeds a certain critical threshold. Then double-diffusive convection rolls develop in thin layers located in regions where the salt stratification is weakest. An important clarification is needed: the curves plotted in figure 3.12 for the density and the buoyancy frequency N are obtained measuring the temperature compensated conductivity, so they depend only on the variations of the salinity content in the water column. However, density variations depend both on the salinity (δS) and the temperature (δT) variations in the form $\delta\rho/\rho_0 = -\alpha\delta T + \beta\delta S$, where α and β are the thermal expansion and haline contraction coefficients respectively. Nevertheless, for a lateral difference of temperature $\Delta T = 10$ K, as it is set in our experiment, the $\delta\rho$ due to temperature is two orders of magnitude less than the $\delta\rho$ due to salinity. From this we can conclude that in the upper/lower mixed layer the value of N is close to zero.

The plot of the time averaged azimuthal velocities for the four measured heights (figure 3.12(c)) shows that the zonal flow is prograde at the surface ($z = 94$ mm), then retrograde at $z = 75$ mm, corresponding to the blue convective region, almost zero in the middle motionless stratified layer ($z = 47$ mm) and again prograde at the lowest measured height ($z = 21$ mm). It is useful to introduce a local version of the Taylor and thermal Rossby numbers (Vincze et al. 2016):

$$Ta(z) = \frac{4\Omega^2\lambda^5(z)N^2(z)}{\nu^2 g\alpha\Delta T}, \tag{3.16}$$

$$Ro_T(z) = \left(\frac{g\alpha\Delta T}{\Omega N(z)L}\right)^2, \tag{3.17}$$

where $\lambda(z)$ is the vertical extent of a convective cell at height z. $\lambda(z)$ is determined by the initial buoyancy frequency profile $N(z)$ and the lateral temperature contrast ΔT, since the thickness of a cell is naturally limited by the condition that the initial (saline) density difference between the top and bottom of the cell cannot exceed the (thermal) horizontal density difference between the lateral sidewalls, as reported by Chen et al. (1971)

$$\lambda(z) = \frac{g\alpha\Delta T}{N^2(z)}. \tag{3.18}$$

More in detail, the procedure used to calculate $\lambda(z)$ is the following.

Let us consider the $\rho(z)$ profile, starting from the top of the water column, i.e. at the surface $z = H$. We want to find the depth z_1 for which it is true that

$$\rho_0\alpha\Delta T = \rho(z_1) - \rho(H), \tag{3.19}$$

i.e. the level where the water density change (due to salinity) from the surface is the same as the density difference due to the ΔT meridional temperature contrast. This level will be the lower boundary of the top cell. So the thickness of this cell is

$$\lambda_1 = H - z_1. \tag{3.20}$$

We can now iterate the same procedure from z_1 downwards

$$\rho_0\alpha\Delta T = \rho(z_2) - \rho(z_1), \tag{3.21}$$

and determine z_2 from the profile. The second cell thickness (from the top) would then be:

$$\lambda_2 = z_1 - z_2. \tag{3.22}$$

As already mentioned, there is a maximum stratification above which the cell formation is blocked and therefore the cells would be too shallow to exist (larger density gradient yields shallower cells). The 'cut-off' N_{critical} for the barostrat experiment is typically reached at

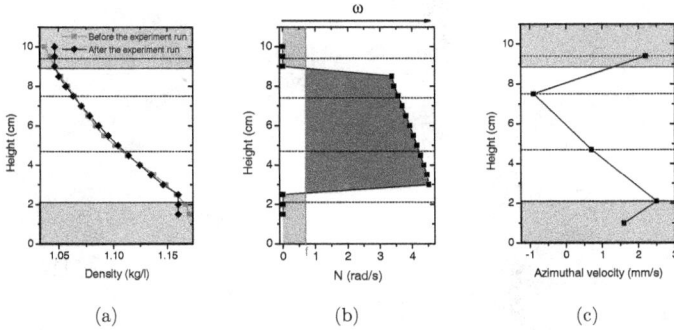

Figure 3.12: Vertical density (a), buoyancy frequency (b), and azimuthal velocity (c) profiles. The dashed lines correspond to the PIV measurement heights, the grey shaded areas indicate the convective layers at the top and the bottom of the tank. (a) the density is measured before starting rotation (red line) and after stopping the rotation (black line). (b) buoyancy frequency calculated from the density measured after the rotation. The vertical red line shows the value of the Coriolis frequency $f = 2\Omega$, the two green areas indicate the frequencies interval in which IGWs can be found ($N < \omega < f$) for the regions where the salinity stratification is weak. The central light blue area indicates the frequencies interval in which IGWs can be found ($f < \omega < N$) for the region with strong salinity stratification. In the latter case, IGWs are expected to propagate similarly to the ones in the atmosphere. (c) plot of the azimuthal velocities averaged over time for the four measured heights. It can be noticed that the zonal flows is prograde at the surface, then retrograde at $z = 75$ mm, corresponding to the convective region, almost zero in the middle motionless stratified layer and again prograde at the lowest z.

mid-depth (Vincze et al. 2016). Whenever $N > N_{\text{critical}}$ there cannot be cells, meaning that there is a region with no convection at mid-depth.

To calculate the thickness of the cell in the lower part of the water column, we repeat the same procedure, but this time starting from the bottom and moving upwards

$$\rho_0 \alpha \Delta T = \rho(0) - \rho(z_b) \qquad (3.23)$$

where $\rho(0)$ is the density at the bottom.

Note that the flow states depend now not only on the lateral temperature difference and on the rotation rate, but also on the convective cell thickness. For this reason, different flow states may be observed at different heights. A detailed discussion about the flow regimes along the water column and inertia-gravity wave signal will be given in chapter 3.2.1.

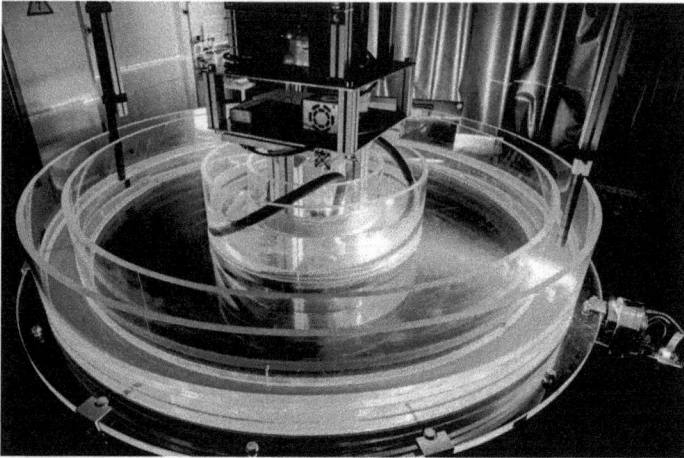

Figure 3.13: Picture of the big differentially heated rotating annulus apparatus. For technical details see table 3.2 and text.

3.3   MSGWs tank

The apparatus described in this section is an experiment newly built at the BTU following the configuration proposed in the numerical simulations by Borchert et al. (2014) with much bigger horizontal dimensions to allow $\Pi < 1$, which we have previously seen is a necessary condition to study IGWs. The experiment has been running from March 2016 and has undergone some modifications and improvements, the most significant of which will also be reported in this section. The investigation of large- and small-scale waves will be presented in detail in chapter 7.

The model of the experiment is the same as the classical annulus previously described. Since the size of the experiment has been increased from the table-size dimensions to much larger-ones, having a diameter of over one meter and a half, some different devices and systems are used (see the picture of the experiment in figure 3.13). The demand for a robust rotating platform, able to carry the weight of several hundreds of kilos (the amount water used for the experiments is itself more than 150 L to which is added the weight of the annulus and the measurement devices placed on the rotating platform), which at the same time needs to have a high precision in the rotation rate and with minima shocks, is a challenge both in terms of engineering requirements and costs.

3.3.1   Rotation unit and water tank

The tank was designed by the technician of the aerodynamic and fluid mechanic department at the BTU (Ludwig Stapfeld, Robin Ströbel, and Vilko Ruoff) and commissioned to the company Heinz-Fritz. The motor support was designed and built at the BTU workshop and afterwards assembled with the motor and the tank. A sketch of the rotation

Geometrical and experimental parameters atmosphere-like annulus			
		plexiglass wall	metallic wall
(a) Geometric dimensions			
Inner radius	a (mm)	350	400
Outer radius	b (mm)	700	700
Gap width	$b - a$ (mm)	350	300
Fluid depth	D (mm)	60	60
(b) Experimental conditions			
T inner wall	T (°C)	12	
T outer wall	T (°C)	35	
Difference of temperature range	ΔT (K)	$0 - 5$	$0 - 10$
Rotation rate range	Ω (rpm)	$0.01 - 2$	$0.01 - 2$
Rotation rate increment	$\Delta\Omega$ (rpm)	0.01	0.01
Ekman layer thickness	δ_E (m)	4.5 to 0.9 $\times 10^{-3}$	4.5 to 2.7 $\times 10^{-3}$
Stewartson layer thickness	δ_s (m)	9.5 to 1.7 $\times 10^{-3}$	2.5×10^{-2}
Thermal layer thickness	δ_T (m)	3.1 to 5.8 $\times 10^{-4}$	4.8×10^{-4}
(c) Fluid's physical properties (de-ionised water)			
Density	ρ (kg m^{-3})	998.21	998.21
Kinematic viscosity	ν (m^{2}s^{-1})	1.004×10^{-6}	1.004×10^{-6}
Thermal conductivity	κ (m^{2}s^{-1})	0.1434×10^{-6}	0.1434×10^{-6}
Expansion coefficient	α (K^{-1})	0.207×10^{-3}	0.207×10^{-3}
Prandtl number	Pr	7	7
Azimuthal wavenumber	m_{min}	3	3
Azimuthal wavenumber	m_{max}	7	8

Table 3.2: Parameters used for the new laboratory experiment with plexiglass inner wall and the modified metallic inner wall.

unit and the water tank is depicted in figure 3.14, where the numbers indicate the main parts described in the following.

1. Metallic base and support of the motor and the tank;

2. synchronous servomotor (model CMPZ100S) with helical-bevel gear unit (KAF57) from SEW-EURODRIVE. The motor has a maximum speed of 3000 rpm and a two-stage gear system provides a high degree of efficiency of over 90% in both torque directions and at all input speeds;

3. rotary index table (series 5) from AUTOROTOR s.r.l. This is a mechanical square axis unit used to transform the uniform rotation of inlet shaft in the rotation of an output disk positioned on top of it;

4. eight metallic arms built to fix the water tank to the rotating disk;

5. the tank consisting of three concentric cylindrical gaps, and made of acrylic glass, or Polymethyl methacrylate (PMMA). The material has been chosen for its features, i.e. light transfer of 92% and specific weight of just 1.19 g cm^{-3}. The acrylic glass resists constant temperatures up to 75 °C. Thanks to the transparent walls and bot-

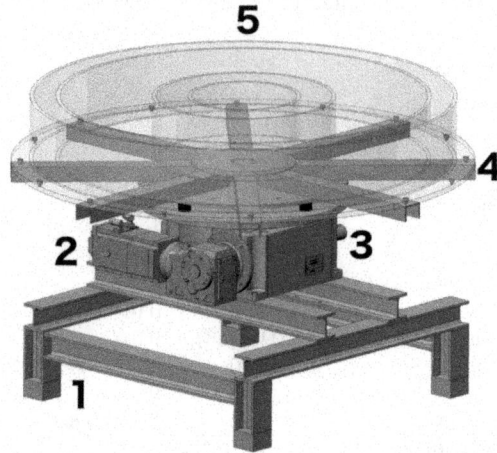

Figure 3.14: Sketch of the motor and the cylindrical tank (concession of Ludwig Stapfeld, Robin Ströbel, and Vilko Ruoff).

tom, the fluid interior is entirely accessible to not invasive measurement techniques such as PIV and infrared thermography (see chapter 4).

3.3.2   Temperature devices

The warm and cold baths in the outer and inner rings are realised with a system of pumps connected to two external thermostats. The refrigerating device is from HAILEA® (HC series chiller model: HC-500A) and it can cool water at any temperature down to 4 °C with a measured accuracy of ±0.05 °C. The heater device is from VAILLANT® (model: VAN 5/6 U plus), and it heatst water up to 85 °C with an accuracy of ±0.1 °C (the accuracy has been tested in the outer ring without rotation). These two devices heat and cool the water at the set temperature (see figure 3.15). In the outer ring, warm water is pumped in and out by two pipes positioned diametrically opposite and fixed in the laboratory frame, i.e. not co-rotating (number 2 and 3 in figure 3.15 respectively). A similar system is used to pump cold water in the innermost cylindrical gap, this time with co-rotating pipes positioned in the water (number 4 in figure 3.15). The middle gap is filled with de-ionised water at the temperature of the lab (kept constant at 21 °C). The rotation of the tank is controlled via a computer (positioned in the middle of the tank and co-rotating, see number 5 in figure 3.15) by software, and the temperature is monitored at the middle of the wall and in the fluid gap, using temperature sensors. Figure 3.16 shows a sketch of the experiment and the position of the temperature sensors is indicated; more details about the experiment geometrical dimensions and relevant quantities are reported in table 3.2.

Figure 3.15: Picture of the tank with the heating and cooling system. The numbers indicates the following components (see text for more details): 1 heating device, 2 and 3 warm water pipes, 4 cold water pipes, and 5 computer.

3.3.3 Temperature difference and heat loss through the walls

Many tests have been conducted in order to study in detail the heat loss through the cylindrical walls of the annulus. Since these walls are made of acrylic glass, which is not an efficient thermal conductor material (thermal conductivity= $0.200 \, \mathrm{W \, m^{-1} \, K^{-1}}$), and they are about 1 cm thick, there is a not negligible heat loss through the walls. A sketch of the experiment with the walls and the position of the temperature sensors used to study the heat loss is depicted in figure 3.16. The water in the outer ring is heated up to $T_1 = 53\,°\mathrm{C}$, and the inner ring is cooled down to $T_6 = 4\,°\mathrm{C}$. However, the difference of temperature measured at the walls in contact with the working fluid in the middle gap is only $\Delta T = T_3 - T_4 \approx 4.5$.

The time series of the temperatures for the six sensors measured during a test run are plotted in figure 3.17. The nomenclature used for the data corresponds to the sensors shown in the sketch 3.16. It can be noticed that after two hours from when the heating and cooling devices are switched on, the temperature in the thermal baths reaches an equilibrium and it stays constant for the entire duration of the experiment. Besides the heat loss mentioned above, another feature can be noticed from the plot: the warm water temperature in the outer ring shows a periodical oscillation with frequency $f = \Omega$, i.e., equal to the rotation rate of the tank. Indeed, before 14:00 the rotation rate was set to $\Omega_1 = 0.1$ rpm and the temperature has oscillations with period $T = 600$ s, whilst after increasing the rotation rate to $\Omega_2 = 0.5$ rpm (this increase is indicated by the vertical magenta line in figure 3.17), the temperature oscillates with a much shorter period corresponding to the new rotation rate. This dependency is most likely due to the fact that the pipes used to circulate the water in the outer ring are not co-rotating on the experiment but fixed in the laboratory frame instead. This, however, should not have a

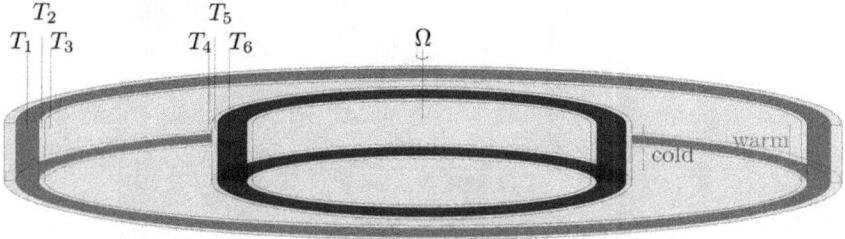

Figure 3.16: Sketch of the rotating annulus having three concentric rings filled with warm water (outermost coloured in red), water at the laboratory temperature (middle gap), and cold water (innermost coloured in blue). Temperature sensors are used and positioned at different radial positions: T_1 outer ring warm water, T_2 middle of the warm wall, T_3 working fluid at the warm wall, T_4 working fluid at the cold wall, T_5 middle fo the cold wall, and T_6 inner ring cold water.

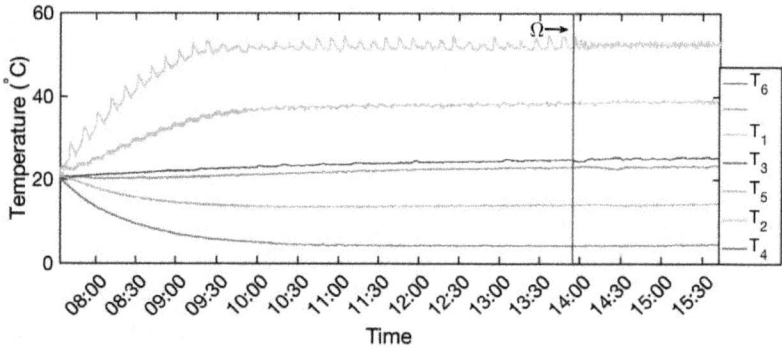

Figure 3.17: Temperature measured at six different radii (see figure 3.16) for the tank with plexiglass wall. Periodic oscillation with a period $T = 600\,\text{s}$, which is equal to 0.1 rpm, i.e. the rotation rate set for the warm-up of the experiment, can be seen in the warm water temperature signal before the rotation rate is increased to $\Omega = 0.5$ rpm.

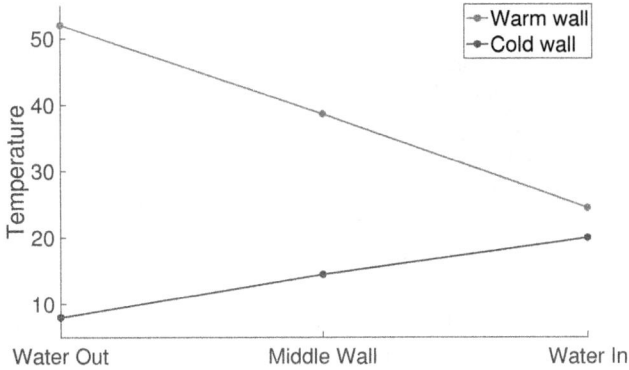

Figure 3.18: Plot of the mean temperature measured by the temperature sensors (see figure 3.16), showing the linear temperature loss due to conduction through the plexiglass walls. The blue data are measured by the sensors T_1, T_2, and T_3 from left to right respectively and the red data are measured T_6, T_5, and T_4 from left to right.

	Water out	Middle wall	Water in
Warm (°C)	$T_1 = 52$	$T_2 = 38.7$	$T_3 = 24.5$
Cold (°C)	$T_6 = 8$	$T_5 = 14.5$	$T_4 = 20$
ΔT (°C)	44	23.5	4.5

Table 3.3: Mean temperature measured by the temperature sensors shown in figure 3.16.

relevant impact on the flow and the waves developing in the experiment, since it can be clearly seen that these temperature oscillations are entirely dumped through the walls and they consequently are negligible at the wall in contact with the working fluid (the temperature oscillations here are less than 1%). The cold water temperature does not show any oscillation related to the rotation rate.

By taking the mean temperature measured by the six sensors when the equilibrium is reached, we can quantify the heat loss. The plot is shown in figure 3.18, and it can be immediately seen that for both walls it follows a linear heat transfer, as it is expected by conduction. The mean temperature values for each sensor are written in table 3.3. The final ΔT at the side of the walls in contact with the working fluid is only 10% of the temperature imposed in the water reservoir.

	Plexiglass wall	Metallic wall
T_W inner ring (°C)	4	9
T cold wall (°C)	19.5	10
T_W outer ring (°C)	53	53
T warm wall (°C)	24.3	19.5
ΔT_{max} (°C)	4.8	9.5

Table 3.4: Comparison fo the measured temperature in the experiments with the plexiglass wall and with the new metallic wall.

3.3.4   Modified configurations

Two main modifications to the original set-up have been done. The first one is a new inner metallic wall added to the tank used to investigate the flow regimes and the small-waves features for higher temperature differences. The second one is a plexiglass lid positioned on top of the working fluid, which has been used to investigate the influence of the upper boundary condition on the large scale flow and small-scale waves, in particular for comparison with numerical simulations.

3.3.4.1   Inner metallic wall

A new inner wall made of black Aluminium with thickness 4 mm has been mounted in the tank. Because the thickness of the wall is $\approx 1/3$ of the original one made of plexiglass and the thermal conductivity of Aluminium is much higher than the conductivity of plexiglass ($237 \, \text{W m}^{-1} \, \text{K}^{-1}$ compared to $0.2 \, \text{W m}^{-1} \, \text{K}^{-1}$), the heat conduction through the new wall is much more efficient. Therefore, a higher lateral difference of temperature can be reached, resulting in a more favourable laboratory set-up to study IGWs. A picture of the experiment with the new wall (in black) is shown in figure 3.19. The first tests showed a decrease in the mean temperature of the working fluid was observed, most likely because the cooling from the inner wall is more effective than the heating from the outer wall. Because of this, the radial temperature difference measured with this new experimental set-up is comparable to the one usually obtained with the original double plexiglass walls set-up.

To overcome this issue and reach a more significant radial temperature difference a different experimental procedure has to be adopted, which is described in the following. The initial rotation is set to $\Omega = 0.4$ rpm during the warm-up time. For the first two hours, only the warm bath is heated, while the cooling bath is kept at a constant temperature $T = 18 \, °\text{C}$. After this first period, also the cooling was set to the final temperature of $T = 6 \, °\text{C}$. After one and a half hour from decreasing the cooling system, the rotation is set to $\Omega = 0.7$ rpm and then increased up to a rotation $\Omega = 1.1$ rpm for which the first baroclinic instabilities can be observed. Right before the baroclinic waves set in, the lateral difference of temperature is $\Delta T = 9.5$; after this, it decreases to $\Delta T \approx 7$. The plot of the temperature measured during the experiment is shown in figure 3.20.

Figure 3.19: Picture of the experiment with the new mounted inner metallic wall (in black).

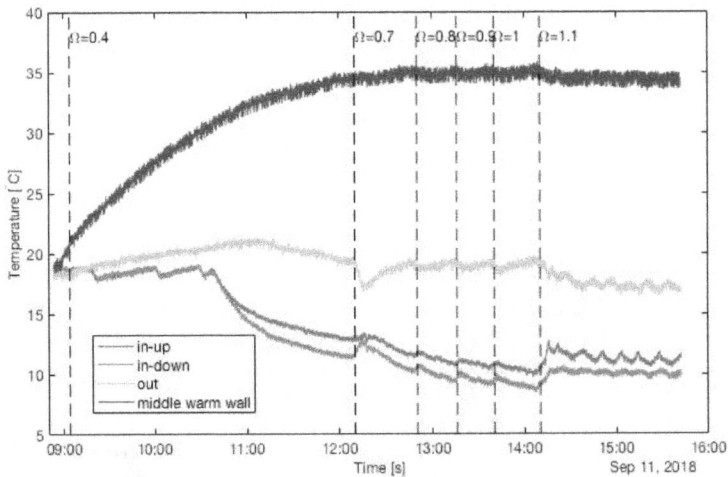

Figure 3.20: Plot of the temperature time serie in the big tank with metallic wall. The data are measured at four locations by the sensor at the middle of the warm wall (violet plot with same name in the legend), by three sensors positioned in the working fluid, one at the warm wall (yellow plot, labelled as 'out') and two at the cold wall (blue and red plots, labelled as 'in-up' and 'in-down').

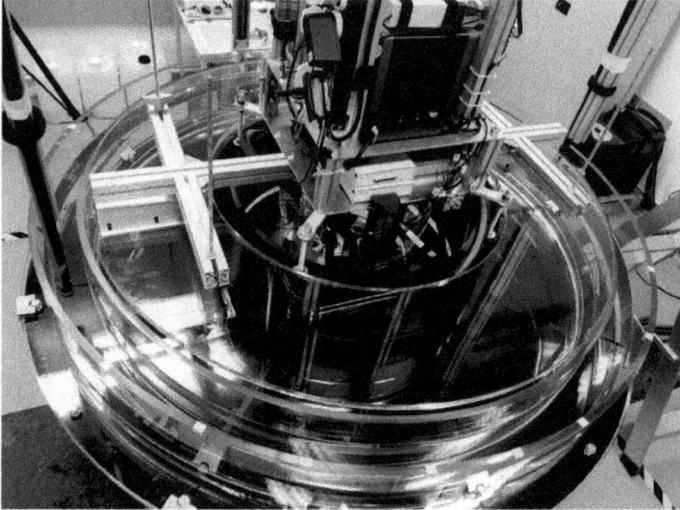

Figure 3.21: Picture of the experiment with the upper plexiglass lid.

	Open surface	Lid immersed	Lid at $1\,$cm from the surface
T_W inner ring (°C)	9		
T cold wall (°C)	10	9.6-10.1	11.6-13.7
T_W outer ring (°C)	53		
T warm wall (°C)	19.5	15.1-16.8	15.4-18
ΔT_{up} (°C)	9.5	6.7	4.9
ΔT_{down} (°C)	9.5	5.5	3.8

Table 3.5: Comparison fo the measured temperature in the experiments with the plexiglass wall and with the new metallic wall. All the measurements are done for rotation rates $\Omega = 1.1$ rpm.

3.3.4.2 Upper lid

The second modification to the experimental apparatus is the application of a transparent lid to cover the water surface. The lid (visible in the picture in figure 3.21) consists of two plexiglass pieces $2\,$cm thick and $29\,$cm wide, covering each half of the surface. The lid is made of the same material as the tank and was commissioned to the same company (Heinz-Fritz). The lid is one centimetre less wide than the gap of the tank with the new wall, in this way we are able to move the lid at different heights, so that we can test different configurations, and some space is left between the walls and the lid where the temperature sensors can be positioned. A small gap (of the order of the millimetre) is also left between the two plates, so that additional sensors can be placed at mid gap (see section 4.1.2 and figure 4.4).

The reason for adding the lid to the experiment is the investigation of the effect of the upper boundary condition on the large-scale flow and small-scale waves. This followed the observation of some discrepancies between the laboratory experiment and numerical simulations performed at the GUF in Frankfurt (see details in section 7.1). Since the numerical simulations neglect the heat exchange between the water and the air above it, the stresses tangential to the surface, surface waves, and the surface tension hints that these phenomena might have an effect on the flow in the annulus.

We will comment on some preliminary results obtained with these modified configurations in section 7.1.4. However, a complete investigation of the dynamics of the waves developing and a comparison with the waves observed in the original configuration is undergoing work and will not be presented in this thesis.

Chapter 4

Measurements techniques

"The grandest discoveries of science have been but the rewards of accurate measurement and patient long-continued labour in the minute sifting of numerical results."

— William Thomson Kelvin —

In this chapter, the measurement techniques used to investigate the flow in the differentially heated rotating annulus are described. The two physical quantities measured are temperature and velocity of the fluid. For each of them, the instruments used are illustrated together with their technical parameters. Since the velocity of the fluid is calculated applying a rather advanced image-processing technique to the measured data, some basics about it are given at the end of the chapter.

4.1 Temperature measurements

Since the fluid in the baroclinic tank is subject to heating and cooling, we can measure the temperature and evaluate its variations within the fluid. In the following sections, we will present the two primary measurement devices used: infrared camera and temperature sensors. The camera is used to investigate 2D surface temperature while the sensor can be placed at different fluid depths and measure the temperature at any given point within the fluid.

4.1.1 Infrared thermography

Infrared thermography is a non-intrusive technique that allows measuring the surface temperature of an object by converting its emanated mid to long-wave infrared radiation. Any solid liquid or gas the temperature of which is larger than the absolute zero emits radiation, as a function of its temperature, because of the continuous motion of its atoms and molecules. The radiation is directly proportional to the temperature: the higher an object's temperature, the more infrared radiation is emitted as a black-body radiation. Infrared energy is just one part of the electromagnetic spectrum, which encompasses radiation from smaller to bigger wavelengths: γ-rays, X-rays, ultraviolet, a thin region of visible light, infrared, microwaves, and radio waves (see figure 4.1). Since the infrared band lies outside the visible region, the radiation cannot be seen by human eyes or standard optical cameras, which detect visible light only.

Frequency [Hz]

10^{20} 10^{18} 10^{16} 10^{14} 10^{12} 10^{10} 10^{8} 10^{6} 10^{4}

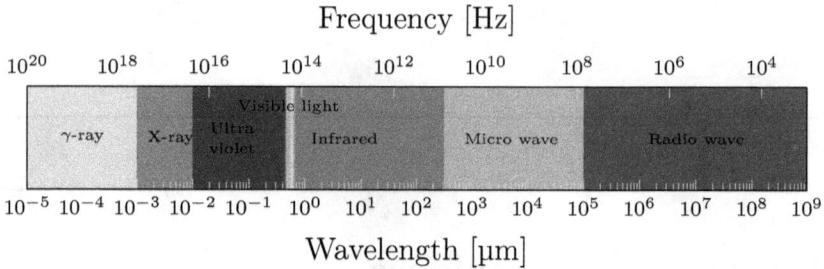

10^{-5} 10^{-4} 10^{-3} 10^{-2} 10^{-1} 10^{0} 10^{1} 10^{2} 10^{3} 10^{4} 10^{5} 10^{6} 10^{7} 10^{8} 10^{9}

Wavelength [µm]

Figure 4.1: Electromagnetic spectrum.

Image format	640×480 pixel
Spectral range	7.5 µm to 14 µm
Range for measuring / visualization	$-40\,°C$ to $300\,°C$
Thermal sensitivity (noise equivalent temperature difference)	$< 80\,mK$
Measurement accuracy	$\pm 1.5\,K$
Dynamic range	16 bit
Image rate	60 Hz (NTSC)

Table 4.1: Infrared camera technical specifications. The infrared camera used is from Jenoptik camera module IR-TCM 640.

A particular camera, called 'infrared camera' or in short IR camera, can detect infrared energy (heat) and convert it into an electronic signal, which is then processed to produce a thermal image on a video monitor and perform temperature calculations. These cameras work even in total darkness because ambient light level, which outside the camera spectral range, does not matter. A critical remark is that glass and plexiglass block long-wave infrared light. Therefore, the focusing lenses of the camera have to be made of special materials such as Germanium or Sapphire crystals, which are much more expensive than ordinary glass. For the same reason, the surface of the fluid must be free or covered by a lid made of the same material.

Most of the currently used IR camera temperature detectors are sensitive in spectral bands with frequencies between 3 and 15 µm, though the band between 5 and 7.5 µm is seldom used because of its rather high atmospheric absorption (Astarita and Carlomagno 2012). Indeed, between the IR camera and the target object, there is an air layer which might absorb part of the radiation. In the frequency window between 7.5 and 13.5 µm the atmospheric absorption is low, and therefore it can be neglected in the IR measurements.

The IR camera used for our experiments is from Jenoptik, model IR-TCM 640. A picture of the camera showing its position in the experimental apparatus can be seen in figure 4.4; the technical specifications are given in table 4.1. The data recorded by the camera are visualised and computer processed via the thermographic software IRBIS® developed by InfraTec. Since the camera is connected via cable to the computer, the temperature maps can be visualised live during the experiment. Time series can be recorded and saved as ASCII files to be then analysed with other data analysis software,

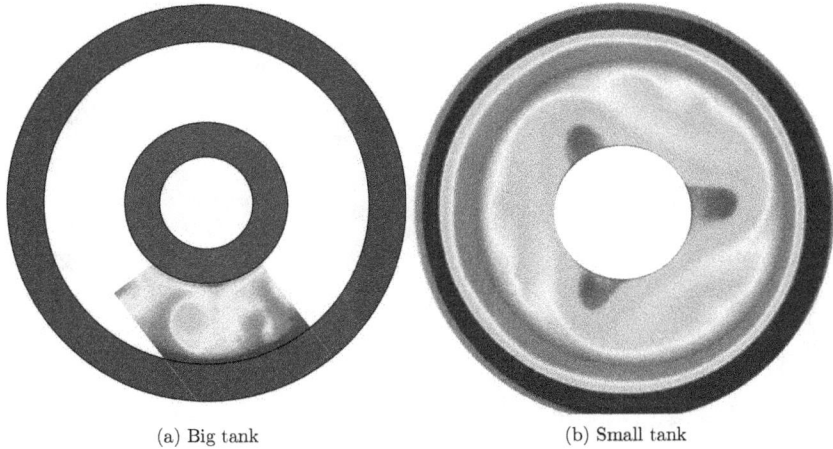

(a) Big tank   (b) Small tank

Figure 4.2: Infrared camera view of (a) the big tank fixed in the lab at $H = 152\,\text{cm}$ from the water surface, (b) small tank fixed or co-rotating at $H = 100\,\text{cm}$ from the water surface.

for example Matlab. For the big tank experiment, the camera is fixed perpendicularly to the water surface in the laboratory frames at a height $H = 152\,\text{cm}$ and has a view of about $1/5$ of the fluid gap (see snapshot in figure 4.2(a)). For steady waves, the entire domain can be reconstructed by superposing snapshots taken at subsequent times (an example of a partially reconstructed image can be seen in figure 4.3(a) and then the complete reconstructed plot in figure 4.3(b)). For the small tank experiment, the camera is similarly fixed perpendicularly to the water surface, at a height $H = 100\,\text{cm}$. For this configuration, the field of view captures the entire experiment (see snapshot in figure 4.2(b)) and there is the option to co-rotate the IR camera.

4.1.2   Temperature sensors

The temperature of the fluid can also be measured at different depths by using temperature sensors. Since the probes are immersed in the fluid, the flow moving over them will be disturbed. Indeed, when the fluid passes over the probe, a boundary layer is formed creating a velocity and temperature gradient. Therefore, in contrast to the IR measurements, this technique is invasive. To minimise the disturbances to the flow, sensors with a minimal diameter are preferably used. Indeed, the object immersed in the fluid will cause friction the magnitude of which is proportional to the object dimensions and the velocity of the flow. A non-dimensional parameter, which measures the ratio between inertia and viscous forces, is the Reynolds number $Re = \rho V D / \mu$, where V is the characteristic velocity of the flow, D the length, and μ the dynamic viscosity. Re can be used to determine the fluid flow conditions. For low Reynolds number, viscous effects are significant enough to damp any disturbances or perturbations in the flow, and the flow remains laminar. Since the flow in our experiment is usually rather slow, small diameter probes should not significantly affect it (Hignett et al. 1985).

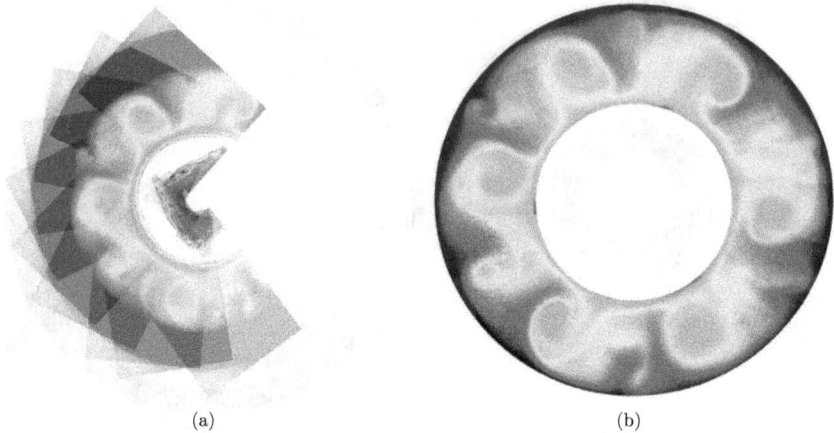

(a)   (b)

Figure 4.3: Reconstruction of the IR camera images for the big tank. Partial reconstruction with 10 snapshots (a), complete reconstruction for wave $m = 6$ (b).

For the big tank, three different types of sensors are used to measure the temperature during our laboratory experiments.

Mid-wall sensors This set consists of sixteen probes (sensor type PT100 Class A from Omega, operating temperature range $-200\,°C$ to $600\,°C$ with accuracy $\pm 0.15\,°C$) placed at mid-thickness of the hot and cold walls (eight sensors for each wall) and equally distributed around the circumference. They are used to check that the walls have a uniform temperature and to measure the heat loss through the walls.

Cold and Warm wall sensors This second set consists of four sensors (sensor type K thermocouple NiCr-NiAl from Omega, operating temperature range $-200\,°C$ to $1250\,°C$, with accuracy $\pm 0.15\,°C$) of diameter $d = 1.4\,mm$ that are attached to the walls on the side of the working fluid (the position is shown at the left-hand side of figure 4.4). They are fixed at two different height, one close to the water surface ($z = 5\,cm$) and one close to the bottom of the tank ($z = 1\,cm$). In this way, the temperature difference between the fluid at the warm wall side and the fluid at the cold wall side can be monitored.

In the fluid sensors The last type is a set of five sensors (thermocouples sheathed sensors consisting of two wires NiCr-Ni from ALMEMO®) of smaller diameter $d = 0.5\,mm$, and the length of the wire is $l = 10\,cm$. Their operating temperature range is $-200\,°C$ to $900\,°C$ with accuracy $\pm 0.1\,°C$. An ALMEMO® precision measuring instrument is connected to the probes and used for measured data acquisition. These devices are used to measure the temperature in different locations in the working fluid (an example of the sensors positioned along a vertical line is shown on the right-hand side of figure 4.4, together with a picture of the sensor and the control unit). The last ones are also used for some of the experiments done

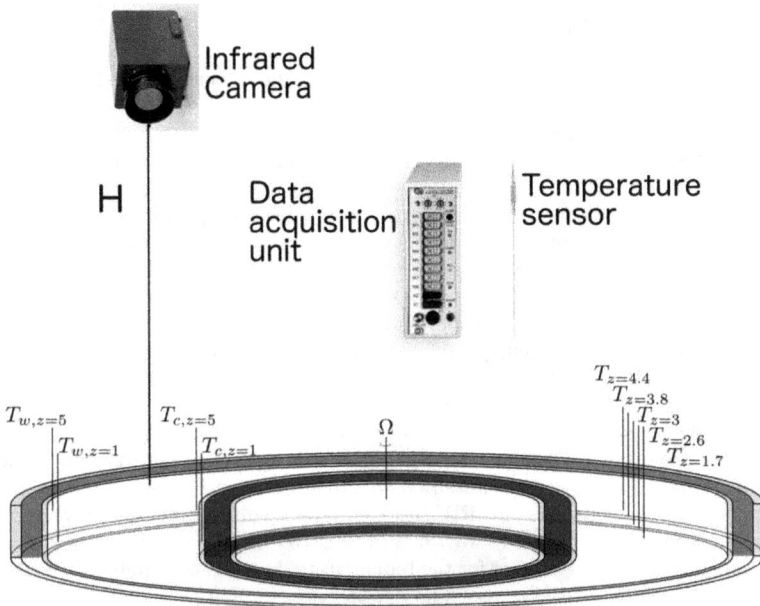

Figure 4.4: Sketch of the big tank with the temperature measurement devices. The infrared camera is fixed in the lab frame at a distance $H = 152\,\mathrm{cm}$ from the water surface. On the left-hand side the position of the sensors used to measure the temperature at the warm and cold walls is displayed. A picture of the ALMEMO® control unit and sensors is shown and an example showing the position of the sensors for one particular experiment is sketched on the right-hand side. See text for the technical specifications of the instruments.

with the small tank (see later chapter 3.2.1). Giving the diameter of the sensors, the Reynolds number is of the order $Re = 1$ and therefore it does not significantly affect the flow. The temperature sensors have a sampling rate $\Delta t = 1$ s and are able to record the data for the entire duration of the experiment.

The first two sets of sensors are used to monitor the radial temperature difference between the warm and the cold walls. The third set is used, instead, for temperature measurements within the flow and the data collected from those is used to investigate the properties of the flow. The accuracy of the three sensor types has been tested in baths set at constant temperatures in ranges variating between $10\,^\circ\mathrm{C}$ and $20\,^\circ\mathrm{C}$ to correct possible systematic errors.

4.2 Velocity measurement

Besides temperature, we are interested in measuring the velocity of the flow to study its properties. To do so, we use the so-called Particle Image Velocimetry, PIV in short, which is a non-intrusive measurement technique.

The concept of PIV is rather simple: neutrally buoyant seeding particles are mixed with the working fluid which is illuminated by a planar laser sheet (for two-dimensional measurements); the illuminated particles are recorded by an optical camera and the flow is reconstructed via a computer imaging system using a tracking method (for a complete review of this method the reader is referred to the textbook by Raffel et al. (2018)). Differently from other methods, PIV does not track the position of each particle, instead it uses ensembles of particles the displacement of which is measured. Since we use planar laser sheet, only two components of the velocity vector can be determined. This method is referred to as 2C-2D-PIV. Some more advanced methods are available to measure the full 3D velocity field but are not used in this thesis.

One of the advantages of using this technique is that, by measuring the displacement and time, it directly gives the fundamental dimensions of the velocity without the need to calculate it indirectly from other measured quantities as other methods, like laser Doppler velocimetry, do. For our system, PIV has the advantage to be non-intrusive so that it does not disturb the flow in any way, it measures two of the velocity components on a plane, similarly to the IR camera for the temperature, but it is not limited to the surface, since the laser can be placed at any fluid depth.

4.2.1 PIV set-up

In our set-up, we position a green laser at the side of the tank mounted on a co-rotating support, which allows us to smoothly move the laser up and down with a small motor actioned by remote control so that we do not disturb the measurements or introduce any shock. The laser beam goes through a lens which splits it into a light sheet of thickness ≈ 1 mm.

A small co-rotating camera is fixed at a certain height above the fluid surface and records the light scattered from the tracing particles (see the sketch in figure 4.5).

4.2.1.1 Camera

The camera used to record the flow is a GoPro hero 4 black, of weight 82 g and dimensions $58 \times 40 \times 28$mm (shown in figure 4.5). The settings of the camera typically used for our experiments are: video resolution 1080p, which corresponds to a screen resolution of 1920×1080, frames per second (fps) 30 or 48, and narrow field of view. Although the camera is capable of higher video resolution (up to 4K) and a higher image rate (maximum 120 fps), we performed several tests which showed that the gain in PIV accuracy using higher resolution videos is minimal compared to the increase in the computation time needed to process the data. The used settings represent the optimal balance for our purposes. A modification that improved the data, was to change the lenses of the camera with some provided by The Imaging Source® (technical specifications in table 4.2). The big advantage of changing the lenses is that we can control the focusing and the aperture, which resulted in being fundamental parameters for the quality of our PIV measurements. Moreover, with the new lenses, we significantly reduce the optical distortion before the

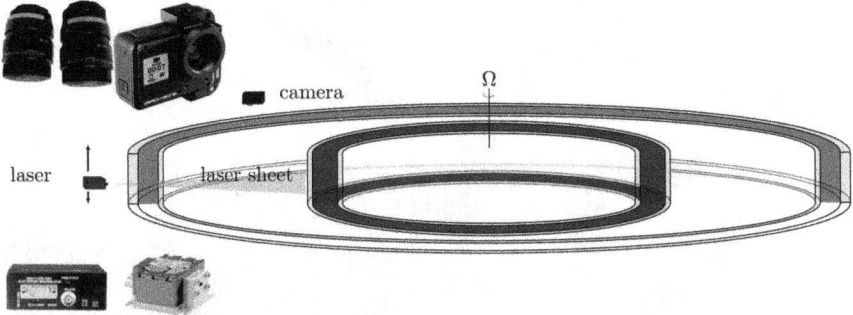

Figure 4.5: PIV experimental set-up with modified GoPro camera and lenses used and Linos nano 250-532-100 laser with its controller unit.

Lens model	Mount	Focal length	Resolution
TCL 1216	C	12 mm	5 MP
TCL 0814	C	8 mm	5 MP

Table 4.2: Lenses The Imaging Source® technical specifications.

data processing. The modified camera has been used for experiments with the big tank (see later chapter 7), whilst the original version has been used for the experiments with the small tank (see later chapter 3.2.1).

4.2.1.2 Laser illumination

Two different lasers are mounted on the two experiments (small- and big-tank). Both are diode-pumped steady lasers emitting a monochromatic beam at wavelength 532 nm. The laser beam goes through a cylindrical-lens optical system that spreads it into a 2D plane sheet of thickness ≈ 1 mm, which is small enough to be approximated to a two dimensional plane. In PIV literature it is often reported the use of double pulses lasers, but for our experiment we opted for steady lasers, which are appropriate for flows with moderate velocity like ours and are much more practical since they do not need to be synchronised with the camera. For the small tank, the model of laser is LLM-SET-50-532 from the company MediaLas. For the big tank, the model of the laser used is Nano 250-532-100 from the company Linos AG. The technical specifications for both are given in table 4.3. The laser illuminates only a sector of the experiment, approximately one third

	Linenlaser LLM-SET-50-532	Linos nano
Max power	50 mW	100 mW
Wave length	532 nm	532 nm
Ray diameter	not specified	1.2 mm
Divergence	not specified	< 0.9 mrad

Table 4.3: Lasers from Linos (used on the big tank) and MediaLas (for the small tank) specifications.

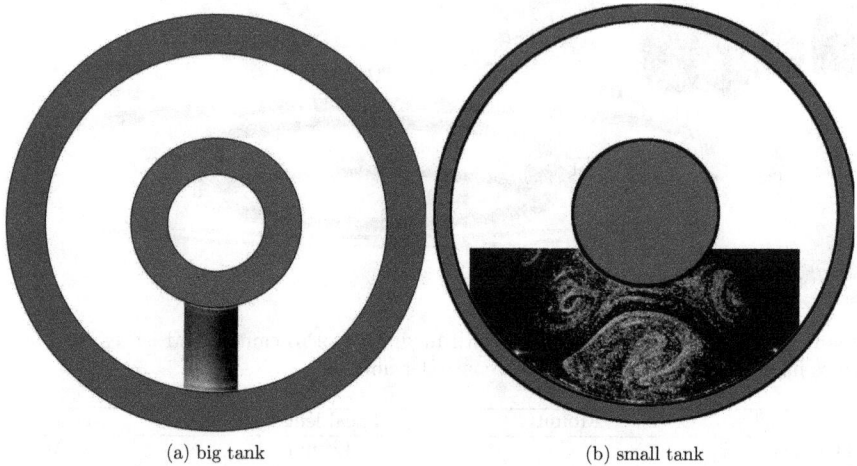

(a) big tank   (b) small tank

Figure 4.6: Picture of the particles in the tank illuminated by the laser (a) big tank and (b) small tank (taken by Dr. Ion Borcia). Note that the two sketches are not in scale.

for the small tank and less than one-sixth for the big tank, as it can be seen in figure 4.6. The PIV measurement technique is mainly used to gain insights on small scale waves; therefore we chose to have high-resolution data on a smaller section of the domain rather than a lower resolution for a larger field of view.

4.2.1.3 PIV particles

Since PIV determines the velocity of the tracer particles and not the one of the fluids itself, we have to make sure to choose appropriate particles whose motion does not significantly depart from the flow. Therefore, besides being sufficiently bright to provide contrasting patterns, the particles must be small enough to minimise inertia and sedimentation effects. In our experiments, we used two different PIV particles, both from DANTEC, namely hollow glass spheres (HGS) and silver-coated hollow-glass spheres (S-HGS). The relevant technical specifications of the two particles types are given in Table 4.4.

For the experiments done with the big-tank, where the working fluid consists of fresh water, the HGS particles are utilised, since their density is closer to the one of the fluid. For the barostrat experiment, since the salt water is denser than the fresh water, we used a mixture of the two particle types.

To calculate the sedimentation time, we can use the Stokes' drag law, a mathematical equation that allows calculating the settling velocities of small particles in a fluid. Under the following simplifying assumptions

1. there is no other particle nearby that would affect the flow pattern

2. the motion of the particle is constant

PIV particles technical specifications	HGS	S-HGS
Mean particle size (μm)	10	10
Size distribution (μm)	2-20	2-20
Particle shape	spherical	spherical
Density (g/cm^3)	1.1	1.4
Refractive index	1.52	-
Material	barosilicate glass	barosilicate glass

Table 4.4: Technical specifications for hollow glass spheres (HGS) and silver-coated hollow-glass spheres (S-HGS) PIV particles used in the laboratory experiments.

3. the particle is spherical and rigid

4. the velocity right at the particle surface is zero

5. the fluid is incompressible

the velocity at which a particle sinks through a liquid column under the influence of gravity can be calculated as

$$V_t = \frac{gd^2(\rho_p - \rho_f)}{18\mu} \tag{4.1}$$

where g is the gravity's acceleration, d the particles diameter, ρ_p the particles density, ρ_f the density of the fluid, and μ is the viscosity of the fluid. The settling velocity and the settling time are proportional to the diameter of the spherical particle squared. The larger the sphere diameter, the faster the particle will settle. The second most critical variable is the difference in the density of the particle and the density of the density of the liquid. This means that matching the density of the particle to the density of the liquid is very important for minimising settling velocity and maximising the time microspheres spend in suspension.

By taking the values reported in table 4.4, we have $V_t = 5.5 \times 10^{-7} \mathrm{m\,s^{-1}}$ for the big tank and $V_t = 1.3 \times 10^{-7} \mathrm{m\,s^{-1}}$ to $1.3 \times 10^{-5} \mathrm{m\,s^{-1}}$ for the barostrat experiment, showing that the particles can be considered in neutral buoyancy during the measurement time typically of the order $T \simeq 10^3$ s. Besides the sedimentation, the convection is strong leading to an effective mixing of the particles.

4.2.2 PIV data processing

In order to obtain fluid velocities, sophisticated image-processing techniques need to be applied to the frames recording by the camera, so that the displacement of the light pattern between two images can be calculated. For this purpose, the free Matlab toolbox UVmat, developed at LEGI in Grenoble, has been used (downloadable at http://servforge. legi.grenoble-inp.fr/projects/soft-uvmat). Useful documentation about the use of the software can be find in Sommeria (2003).

The software contains a Matlab program for PIV the Correlation Image Velocimetry (CIV) developed by Fincham and Spedding (1997) and Scarano and Riethmuller (2000). This technique relies on generalised pattern matching by direct cross-correlation of the luminous intensity between image pairs, limited to small subdomains. The velocity vector

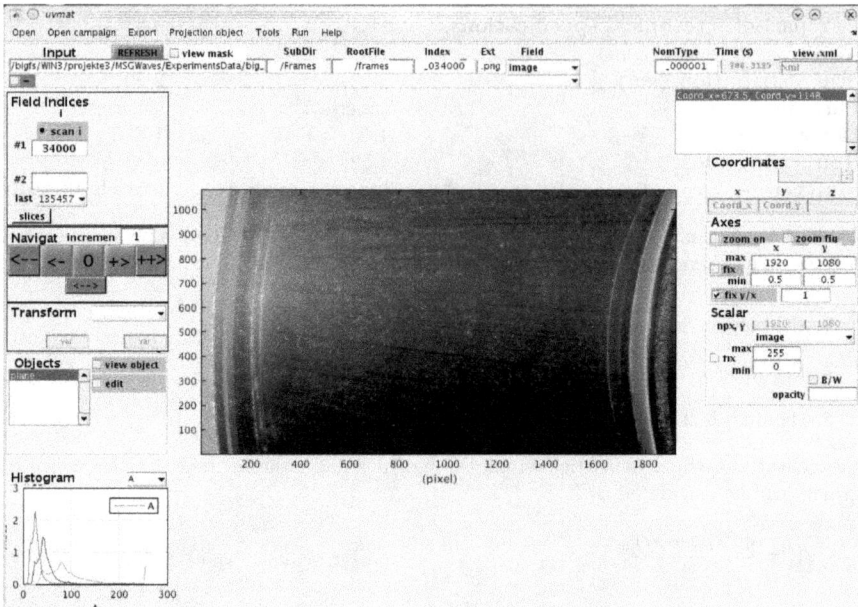

Figure 4.7: Main UVmat display window showing a frame recorded in the big tank experiment.

on each box is determined as the displacement which maximises the covariance, multiplied by the time between the images. The software includes iterative CIV, taking into account pattern deformation by strain and rotation.

The procedure for getting the velocity field from the recorded video is the following. First of all the video is converted to a series of frames using the ffmpeg toolbox. Subsequently, the image is imported in Matlab with the UVmat toolbox and visualised in a graphic window (see figure 4.7). An appropriate mask can be drawn and applied to the image to select only the areas of interest.

Then, the user has to set the appropriate parameters from a user-friendly interface visible in figure 4.8. The window consists of seven rows, corresponding to the main operations done by the software during the data processing.

0. Import the frames and choose time interval

1. first iteration

   • civ1: initial image correlation process

   • fix1: removal of false velocity vectors

   • patch1: filtering and interpolation on a regular grid

2. second iteration

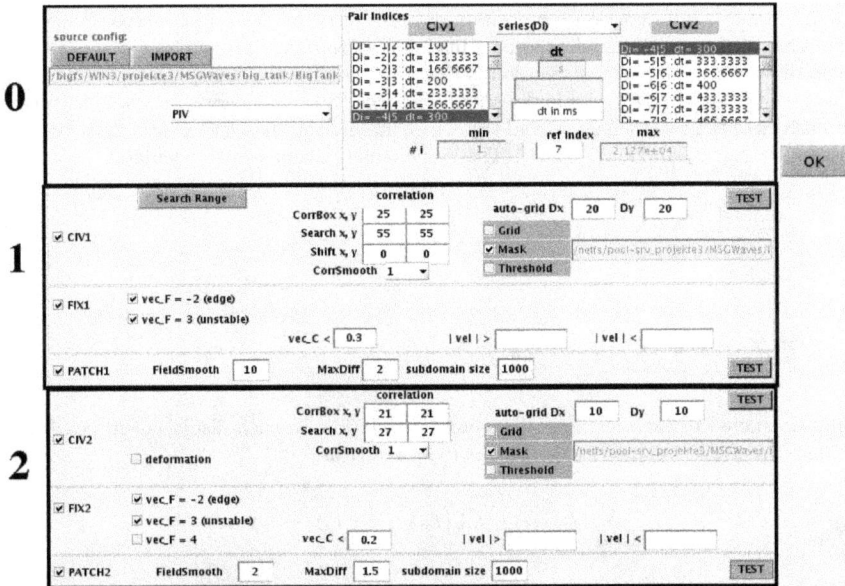

Figure 4.8: Snapshot of the UVmat configuration used to calculate the horizontal velocities from the recorded movies. The three black boxes show the initial operation of choosing the pair of images (0), the first iteration (1), and the second iteration (2).

- civ2: advanced image correlation (using results form the first iteration as inputs)
- fix2: removal of false velocity vectors
- patch2: filtering and interpolation on a regular grid

In the first step, two images separated by a time interval dt are loaded. In the upper right corner, there are two boxes which allow choosing the dt; it is recommended to select it such that the displacement of the seeding particles covers $\approx 5-20$ pixels. It is clear that the analysis of various flows having different typical velocities requires to assign to each a proper dt; therefore the parameters have to be evaluated for each experiment separately, and there is no standard configuration. UVmat provides the possibility to visualise the two images at t and $t+dt$ in a rapid sequence so that it is easier to evaluate the appropriate dt. The second step, second line in figure 4.8, is **CIV1**, which stands for correlation imaging velocimetry. This part is the core of the data processing. The user is asked to set the size of the correlation box (CorrBox x,y,) and the searching box (Search x,y) both in pixels unit. The concept of CIV is shown in figure 4.9 where the correlation box is indicated in the image (a) with size (Bx,By). This box defines the region which cross-correlation between the previously selected image pair will be calculated along each direction (x, y). The size of the correlation box should be sufficiently large to contain at least five particles but small enough in order to have a good spatial resolution. It is, therefore, important to have a dense and uniform seeding particle density. The size of the search box should be adequately big to contain the position of the tracked particles in the second image. If the search range is too small, the true maximum covariance cannot be found, and false vectors pop up. The cross-correlation is then calculated with all the possible positions of the pattern box (Bx,By), inside the search box (Sx, Sy). The algorithm works with pixel intensities of the two images (a) and (b), which are indicated as $I_a(k,l)$ and $I_b(k+i,l+j)$, with $k = 1,...,B_x$ and $l = 1,...,B_y$. The box intensity averages are calculated

$$\overline{I_a} = \frac{1}{B_x B_y} \sum_{k=1}^{B_x} \sum_{l=1}^{B_y} I_a(k,l), \tag{4.2}$$

$$\overline{I_b} = \frac{1}{B_x B_y} \sum_{k=1}^{B_x} \sum_{l=1}^{B_y} I_b(k+i,l+j), \tag{4.3}$$

and subtracted to each intensity. The cross-correlation for each displacement $c(i,j)$ normalised by the covariance is then calculated directly as

$$c(i,j) = \frac{\sum_{k=1}^{B_x} \sum_{l=1}^{B_y} \left(I_a(k,l) - \overline{I_a}\right)\left(I_b(k+i,l+j) - \overline{I_b}\right)}{\left[\sum_{k=1}^{B_x} \sum_{l=1}^{B_y} \left(I_a(k,l) - \overline{I_a}\right)^2 \sum_{k=1}^{B_x} \sum_{l=1}^{B_y} \left(I_b(k+i,l+j) - \overline{I_b}\right)^2\right]^{\frac{1}{2}}}. \tag{4.4}$$

The velocity vector is the displacement that maximises the covariance $c(i,j)$ divided by the time interval. The covariance is interpolated to not integer displacements, and therefore the precision obtained is of the order of sub-pixel.

By directly computing the correlation, instead of using Fourier transforms as other PIV software do, UVmat is more demanding in terms of computational costs, but it is more precise and allows smaller research boxes, which means it has a higher resolution.

The third step **FIX1** is a routine that removes the vectors with low correlation, according to a threshold (vec_C <) set by the user.

The fourth step **PATCH1** interpolates the vectors on a regular grid and applies a smoothing filter. This prepares the data for the next iteration of the program.

In the second iteration, the three steps (CIV, FIX, and PATCH) are repeated, but this time the input file is the results obtained by the first iteration. Since the second iteration is based on the previously calculated displacement, the spatial resolution in this step can be improved by reducing the pattern-box size, and a better precision can be achieved by choosing a larger time interval. After obtaining the velocity vectors, the last operation needed is to convert the coordinates from pixels to physical coordinates. To do so, a calibration grid covering the entire field of view is used (see figure 4.10 (3)). Once the dimensions of the grid are known and inputted in UVmat, which can then determine the intrinsic and extrinsic parameters of the camera and from that the transformation between the two coordinate systems is done. During this operation, camera distortion is also compensated. Once the calibration is completed, the velocity, divergence, and curl data are available for analysis.

The main steps necessary to process the data are summarised in figure 4.10.

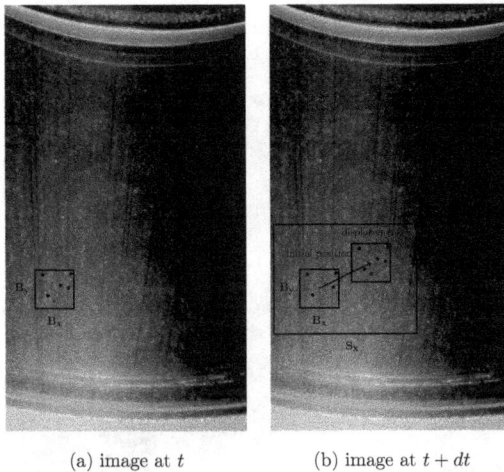

(a) image at t (b) image at $t + dt$

Figure 4.9: Sketch of CIV between to images at times t (a) and $t + dt$ (b). In (a), the correlation box of dimensions B_x and B_y is sketched. Note that the boxes size has been enlarged for showing purposes. In (b), the search box of dimensions S_x and S_y is sketched, showing the displacement of the tracked particles (indicated by the arrow) from their initial position.

For a complete overview of all the functions and options of the software, the reader is referred to the UVmat documentation available on the webpage: `http://servforge.legi.grenoble-inp.fr/projects/soft-uvmat/wiki/UVmatHelp`.

Compared to other available PIV post-processing software, UVmat offers the possibility to visualise various diagnostic quantities that help the user to check the quality and possible sources of errors at each of the main steps, so it is easy to compare different configurations and find an optimal one by tuning the parameters chosen.

4.2.2.1 PIV error estimation

We have seen in the previous sections how PIV measurements and image-processing techniques involve many steps, during which different sources can play a role in the error associated to the velocity. Because of the variety of error sources and the difficulty to quantify them separately, a global estimation of the PIV uncertainty can be challenging.

A first estimation of the error coming from the PIV software used to get the velocities field from the images (UVmat) is given by two quantities directly estimated by the software, namely the rms difference between the CIV (Correlation Image Velocimetry) and the smoothed velocity field, and secondly the number of vectors excluded (`http://servforge.legi.grenoble-inp.fr/projects/soft-uvmat/wiki/Tutorial/CorrelationImageVelocimetryOptimisation`). Figure 4.11 shows an example of the error estimation provided by UVmat.

Another possibility is to use an *a posteriori* error estimation technique that allow us to take into account mostly of the error sources and give a quantitative evaluation of the

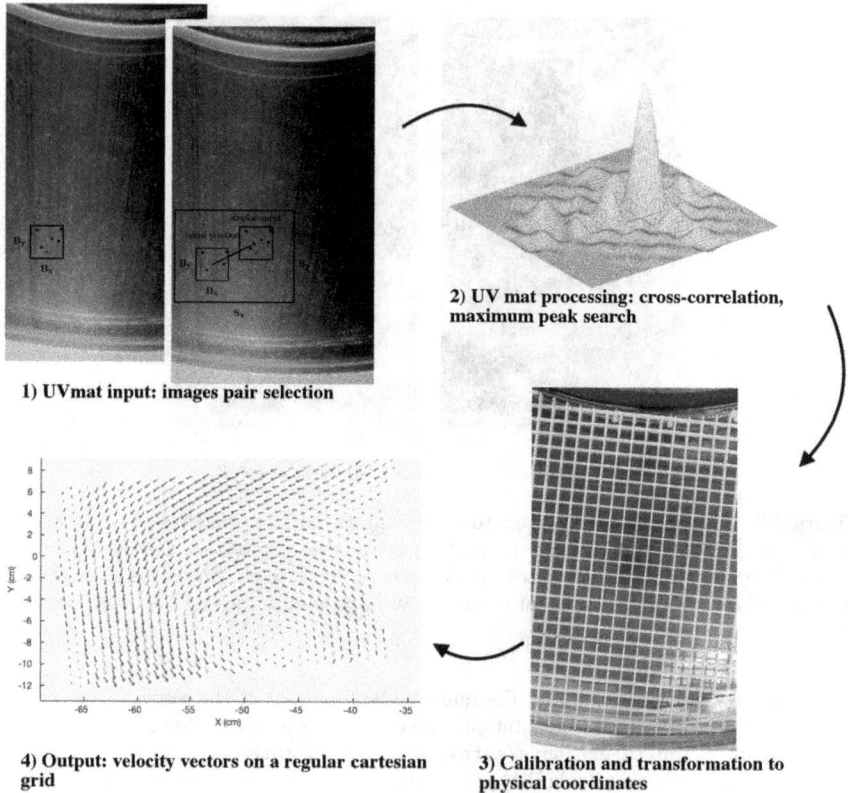

1) UVmat input: images pair selection

2) UV mat processing: cross-correlation, maximum peak search

4) Output: velocity vectors on a regular cartesian grid

3) Calibration and transformation to physical coordinates

Figure 4.10: Schematic overview of the main operations done with UVmat to obtain the velocity field from the raw images.

measurement error from the processed data directly. From this idea, a second way to quantify the error follows. This second method consists of taking two consecutive PIV images ($\Delta T = 0.03$ s) and subtracting the velocities field. Because our flow is rather slow, one would expect that the fields do not differ too much and they are completely uncorrelated one from the other because to compute the PIV we use two different sets of images. From this we can calculate the relative error for the velocity components as:

$$err = \frac{V_1 - V_2}{V_1 + V_2} \tag{4.5}$$

Repeating this procedure for the entire time serie we can estimate the mean error associated to the PIV data analysis for the U and V components of the velocity. Two plots showing the error obtained for measurements done with the small tank in the barostrat experiment are shown in figure 4.12. In this case, it can be seen how the error uniform over the entire domain and always smaller than 15%. This method is particularly useful

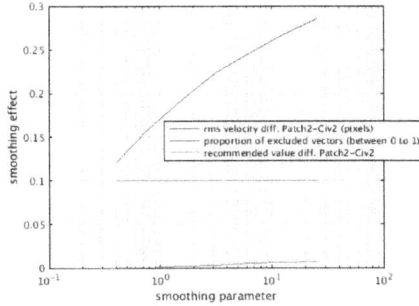

Figure 4.11: Test of the error provided by UVmat.

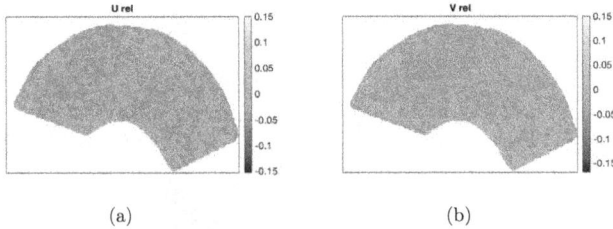

(a)   (b)

Figure 4.12: Relative error for the U (a) and V (b) component of the velocity calculated using (4.5).

to check whether there are particular areas in the domain where errors are bigger than the rest of the domain, as it might be the case when the illumination of the laser sheet is not uniform or the seeding particles are not dense enough or well distributed.

4.2.2.2 PIV resolution

To calculate the resolution of the data after the PIV post-processing, we have to consider the image resolution (X_{res}, Y_{res}) in pixels and divide it by the dimension of the correlation box (B_x, B_y) used in UVmat divided by a factor two, which takes into account a 50% overlap of the boxes. The resolution in pixel will then be

$$N_x = \frac{X_{res}}{B_x/2}, \quad N_y = \frac{Y_{res}}{B_y/2}. \tag{4.6}$$

The resolution of the camera is $X_{res} = 1920$ and $Y_{res} = 1080$, while the usual box size is $B_x = B_y = 20$. These values give $N_x = 192$ and $N_y = 108$. To convert the points into physical units we need to consider the X and Y sizes in centimetres. For the big tank $X = 35$ cm and $Y = 20$ cm, which results in 1point ≈ 0.2 cm. For the small tank we have $X = 25$ cm and $Y = 12$ cm, which results in 1point ≈ 0.13 cm

Chapter 5

Data analysis

In this chapter we describe the methods used to analyse velocity fields obtained via the PIV measurements.

Section 5.1 focuses on the large scale flow, and more particularly on methods devoted to the determination of the different dynamics of the baroclinic waves and investigation of possible interaction with other wave types, such as Kelvin waves.

The following section 5.2 illustrates the different methods used in this thesis to identify IGWs and investigate their properties.

5.1 Data analysis methods for large scale flow

In this first section we focus on the two statistical methods used to analyse the large scale flows and their dynamics. These methods shall be applied to the the barostrat experiment, in which it is interesting to study whether the baroclinic dynamics is different in the two unstable layers and whether there are interactions with other large scale waves that formed, like the observed Kelvin waves (see chapter 6 for the results).

The harmonic analysis is a form of signal demodulation in which the user specifies the frequencies to be examined and applies least square techniques to solve for the constituents (for an extensive description of the method we refer to the textbook of Thomson and Emery (2001)). This method is very useful to investigate the spatial patterns associated with a single given frequency and in particular we use it to emphasise the wave modes at different heights. Moreover, it has the advantage to be very robust even for short time series.

Besides harmonic analysis, we apply the empirical orthogonal function (EOF) analysis to the PIV data (a complete description of the method can be found in the textbook of Navarra and Simoncini (2010)). Without specifying particular frequencies in advance this method provides a description of the spatial patterns of variability of the data series and their temporal variation, breaking the data into orthogonal functions or 'modes of variability' and thus is widely used in geosciences (Lorenz 1956). One mode of variability can contain more than one frequency and therefore can comprise a more complex dynamics related e.g. to interactions between waves. The advantage of using independently both methods is that we can investigate time dependent patterns (with the EOF analysis, whereas the harmonic analysis gives only the spatial patterns) and give them a physical interpretation by comparing the results obtained independently from the two techniques.

Indeed, while the interpretation of the spatial patterns obtained by the harmonic analysis is clear, connecting the EOFs to the physical modes is not straight forward.

The emission of inertia-gravity wave packets from the baroclinic jet is a phenomena highly localised in space and time (Viúdez and Dritschel 2006). Therefore, it is very difficult to capture these small scale waves using the two statistical methods described in this section. A quantity often used as an indicator for IGWs is the horizontal velocity divergence $\nabla_{\mathbf{h}} \cdot \mathbf{u} = \partial u/\partial x + \partial v/\partial y$ (O'Sullivan and Dunkerton 1995, Borchert et al. 2014). The horizontal divergence contains a balanced part, as defined by quasi-geostrophic balance, and an imbalanced part that is related to IGWs. We use the horizontal divergence to study the small scale waves, their behaviour in time, their wavenumbers and frequencies. The latest are obtained by computing two dimensional fast Fourier transforms.

5.1.1   Harmonic analysis

We consider the velocity field V, measured by PIV at an arbitrary grid point in our measured domain $V(t_n)$ at times $n = 0, 1, ..., T$. This quantity can be expressed by a Fourier expansion

$$V(t_n) \approx \overline{V} + \sum_{q=1}^{M} [A_q \cos(2\pi f_q t_n) + B_q \sin(2\pi f_q t_n)] + V_r(t_n) \tag{5.1}$$

with $q = 0, 1, ..., M$ and M is the number of distinct frequencies to be analysed, $\overline{V}$ is the temporal mean, V_r the residual of the time series (it could contain other kinds of components), $t_n = n\Delta t$ the time, f_q a constant frequency, A_q and B_q are the harmonic coefficients of the Fourier series. The amplitude C_q and phase ϕ_q of the frequency component q are given by

$$C_q = (A_q^2 + B_q^2)^{1/2}, \tag{5.2}$$

and

$$\phi_q = \tan^{-1}(B_q/A_q) \tag{5.3}$$

respectively. The goal of the harmonic analysis is to determine the coefficients A_q and B_q, $q = 0, 1, ..., M$, for the M distinct frequencies. The frequencies to be analysed are chosen as the main peaks in the horizontal velocity spectra at each measured fluid height. The horizontal velocity components, measured along one radial line taken in the middle of the camera field of view, are selected. The frequency spectrum for each of these points is then calculated by using a fast Fourier transform algorithm and then the spectrum is averaged for the points of this particular chosen line.

A truncated Fourier series is then fitted to the time series containing the frequencies to be analysed, $\omega_q = 2\pi f_q$. The variance e^2 is computed for each point as

$$e^2 = \sum_{n=1}^{T} \left\{ V(t_n) - \left[\overline{V} + \sum_{q=1}^{M} [A_q \cos(\omega_q t_n) + B_q \sin(\omega_q t_n)] \right] \right\}^2 . \tag{5.4}$$

We estimate A_q and B_q and hence the amplitudes, C_q, and phases, ϕ_q, of the various components by minimising the variance. Once the amplitudes and the phase for a certain frequency are calculated for each grid point, the corresponding velocity field can be plot on the domain recorded by the camera (approximately one third of the tank). For rather steady waves, we can then graphically reconstruct the entire annulus using symmetric properties. The reconstructed plots are made by combining together 3 or 4 partial plots obtained from the harmonic analysis. These are shifted by a phase $\phi = 120°$, when 3 images are used or $\phi = 90°$ when 4 images are used. The figures have then been combined together (with some overlapping) to reconstruct the full annulus, relying on the hypothesis of patterns regular in space. The harmonic analysis is applied to the data from the barostrat experiment in section 6.2, and examples of reconstructed images can be seen, for example, in figure 6.4.

5.1.2 Empirical orthogonal functions

To analyse data that contain a wave-like propagating signal, it is useful to use a modified version of the standard EOF analysis, the so called complex empirical orthogonal functions (CEOFs) (Pfeffer et al. 1990).

For a harmonic wave of the form $V(x,t) = Re[U(x)\exp(-i\omega t)]$ a peculiar phase relation that indicates propagation is a quarter wavelength shift. CEOF analysis enhances this phase relation changing the available data by adding a new data set obtained by shifting all data by one quarter wavelength using a Hilbert transform (Navarra and Simoncini 2010). Therefore, a single CEOF represents a single mode split into two patterns with a phase difference of $\pi/2$ (the real and imaginary part of the CEOF).

To calculate the CEOFs we follow the same approach described in Harlander et al. (2011), considering the CEOFs method for a simultaneous analysis of more than one field. Such a coupled analysis is useful in our case where one field is the u- and the other is the v-component of the velocity. We proceed in the following way to find coupled propagating patterns: complex time series of the velocity components are formed from the original time series and their Hilbert transforms:

$$u_c(x,t) = u(x,t) + iu_H(x,t), \tag{5.5}$$

$$v_c(x,t) = v(x,t) + iv_H(x,t), \tag{5.6}$$

where $u(x,t)$, $v(x,t)$ are the time series of the horizontal components of the velocity measured by PIV at each location in the recorded domain, $u_H(x,t)$ and $v_H(x,t)$ are the Hilbert transforms of $u(x,t)$, $v(x,t)$ and i is the imaginary unit.

The Hilbert transform is defined as:

$$X_t^H = \sum_\omega \zeta^H(\omega)\exp(-2\pi i\omega t), \tag{5.7}$$

of the original time series X_t with Fourier decomposition

$$X_t = \sum_\omega \zeta(\omega)\exp(-2\pi i\omega t), \tag{5.8}$$

where $\zeta^H(\omega) = i\zeta(\omega)$ for $\omega \le 0$ and $\zeta^H(\omega) = -i\zeta(\omega)$ for $\omega > 0$. For the computation of the Hilbert transforms, we used the algorithm described by Marple (1999).

We then form extended time series by combining the $v_c(x,t)$ time series with the $u_c(x,t)$ time series and rewriting them as a row-vector in the form of $U = (u_1, ..., u_M, v_1, ..., v_M)$. The data matrix D is written in the form of

$$D = \begin{pmatrix} U_1(t_0) & U_2(t_0) & \dots & U_M(t_0) \\ U_1(t_1) & U_2(t_1) & \dots & U_M(t_1) \\ \vdots & \vdots & & \vdots \\ U_1(t_N) & U_2(t_N) & \dots & U_M(t_N) \end{pmatrix} \tag{5.9}$$

where the rows represent the state vector at the spatial grid points $U(t_n) = (U_1(t_n), ..., U_M(t_n))$ at time t_n and the columns represent the time series $U_m(t) = (U_m(t_0, ..., t_N))^T$ at the spatial point m.
From the data matrix, we calculate the covariance matrix

$$F = D^T D, \tag{5.10}$$

The CEOFs are the eigenvectors of F and the so called Principal Components (PCs) are the corresponding time-dependent coefficients. These are obtained by projecting for each time step the EOF on the data

5.2   Data analysis methods for IGWs

Unbalanced motion and small-scale phenomena are usually embedded into the large-scale balanced flow. Therefore, the first step towards studying internal gravity waves from the data measured in the laboratory is to identify them and separate them from other fluid motion. The two statistical methods previously described are not suited to capture these small scale waves because of their high localisation in space and time (Viúdez and Dritschel 2006)

5.2.1   Large-scale/small-scale spatial separation

A quantity often used as indicator for IGWs is the horizontal velocity divergence $\nabla_h \cdot \mathbf{u} = \partial u/\partial x + \partial v/\partial y$ (O'Sullivan and Dunkerton 1995, Borchert et al. 2014). This quantity is useful if we assume that the large-scale flows are mainly divergent-free and therefore small-scale gravity waves can be detected in the divergence field. With this assumption, by calculating the divergence and the curl of the total flow, we can separate the two motions.

The horizontal divergence is formed by a balanced part, as defined by quasi-geostrophic balance, and an imbalanced part that is related to IGWs. Assuming that the balanced part is characterised by large scale signals and that unbalance, instead, is characterised by small-scale signals, we decompose the flow variables in the following way:

$$\mathbf{V}(\mathbf{x},t) = \overline{\mathbf{V}}(\mathbf{x},t) + \mathbf{V}'(\mathbf{x},t) \tag{5.11}$$

where $\mathbf{V}'$ corresponds to the fluctuations and $\overline{\mathbf{V}}$ to the variable filtered over space. To compute the averaged quantities we perform a moving average using a square window. The fluctuations are then obtained subtracting the spatial average from the original data.

5.2.2 Horizontal deformation

As we have seen in section 2.6.2, wave capture is a mechanism that can strongly affect the propagation of inertia-gravity waves. To investigate possible regions of the flow where wave capture can occur, we need to calculate the horizontal fluid deformation, which consists of stretching and shearing deformation.

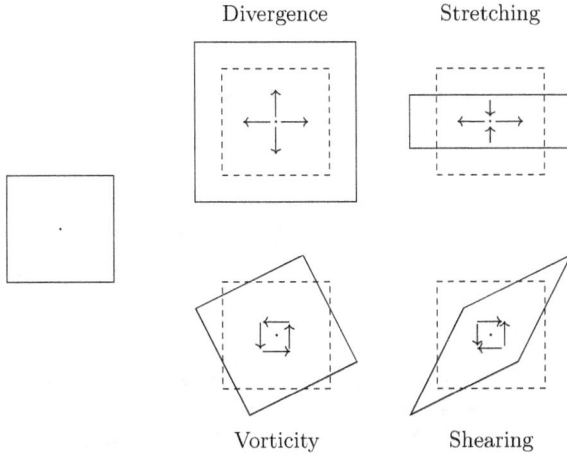

Figure 5.1: Sketch showing the effects of divergence and stretching deformation (top row) and vorticity and shearing deformation (bottom row). Redrawn after Saucier (1953).

In the subsequent description we follow the approach used by Saucier (1953):

The stretching deformation, which implies a change of shape resulting in a stretch along the stretching axis and a contraction along the perpendicular axis, is expressed as

$$a_{\text{stretch}} = \left(\frac{\partial u}{\partial x} - \frac{\partial v}{\partial y} \right). \tag{5.12}$$

The shearing deformation occurs at an angle of $\pi/2$ from the stretching deformation and can be written as

$$a_{\text{shear}} = \left(\frac{\partial v}{\partial x} + \frac{\partial u}{\partial y} \right). \tag{5.13}$$

The resultant deformation is given by the combination of the two components

$$a = \frac{1}{2} \sqrt{a_{\text{stretch}}^2 + a_{\text{shear}}^2} \tag{5.14}$$

This formulation of the deformation, however, it is not very practical since we wish to know the direction of dilatation together with its magnitude. To obtain these informations, one can apply a rotation transformation to the axis. In this new coordinate system either a_{stretch} or a_{shear} vanishes.

By this transformation and with the help of some trigonometric properties (the reader can find the complete derivation in Saucier (1953)) the resulting angle of rotation that sets $a_{\text{shear}} = 0$ is

$$\theta_{\text{def}} = \frac{1}{2} \tan^{-1} \frac{a_{\text{shear}}}{a_{\text{stretch}}}. \tag{5.15}$$

The new coordinate system is rotated of an angle θ_{def} with respect to the original one and its $x - y$ axes correspond to the dilatation and contraction axis. To determine which one is which we have to look at the sign of the shearing deformation. If $a_{\text{shear}} > 0$, x is the dilatation and y the contraction axis, if $a_{\text{shear}} < 0$, y is the dilatation and x the contraction axis. This analysis shall be used in section 7.2.5, where wave capture is discussed.

5.2.3   Spectra

Once wavelike features have been observed and isolated in the physical space, a frequency analysis in the Fourier space is necessary to identify the properties of the waves.

5.2.3.1   Two dimensional Fourier transform

The two dimensional Fourier transform is a generalisation to higher dimensions of the commonly used one dimensional transformation. In one dimension the Fourier transform provides a one-to-one transform of signals from a time-domain representation $h(t)$ to a frequency domain representation $H(\nu)$. This transformation is based on the hypothesis that any function $h(t)$ with $-\infty < t < \infty$ can be expressed as an integral sum of sines and cosines of continuous frequencies in the interval $-\infty < \nu < \infty$. More specifically, the definition of 1D Fourier transform reads

$$H(\nu) = \frac{1}{2\pi} \int_{-\infty}^{\infty} h(t) e^{(i\nu t)} dt, \tag{5.16}$$

and the inverse Fourier transform is

$$h(t) = \int_{-\infty}^{\infty} H(\nu) e^{(-i\nu t)} d\nu. \tag{5.17}$$

The extension of these transformations to the two-dimensional case is straightforward. For time-space data $g(x,t)$ (the plot of which is defined as Hovmöller plot) the 2D Fourier transform is

$$G(\nu, k) = \frac{1}{2\pi} \int_{x=-\infty}^{\infty} \int_{t=-\infty}^{\infty} g(x,t) e^{i(\nu t + kx)} dx dt, \tag{5.18}$$

and the inverse Fourier transform is

$$g(x,t) = \int_{k=-\infty}^{\infty} \int_{\nu=-\infty}^{\infty} G(\nu, k) e^{-i(\nu t + kx)} d\nu dk. \tag{5.19}$$

Since our data are discrete, we use the discrete fast Fourier transform. In this case, the integrals are substituted by summations in the respective equations

$$G_{k,m} = \frac{1}{XT} \sum_{n=0}^{X-1} \sum_{j=0}^{T-1} g_{n,j} e^{(2\pi i \frac{kj}{T})} e^{(2\pi i \frac{mn}{X})}, \tag{5.20}$$

For our data analysis we used the MATLAB function:

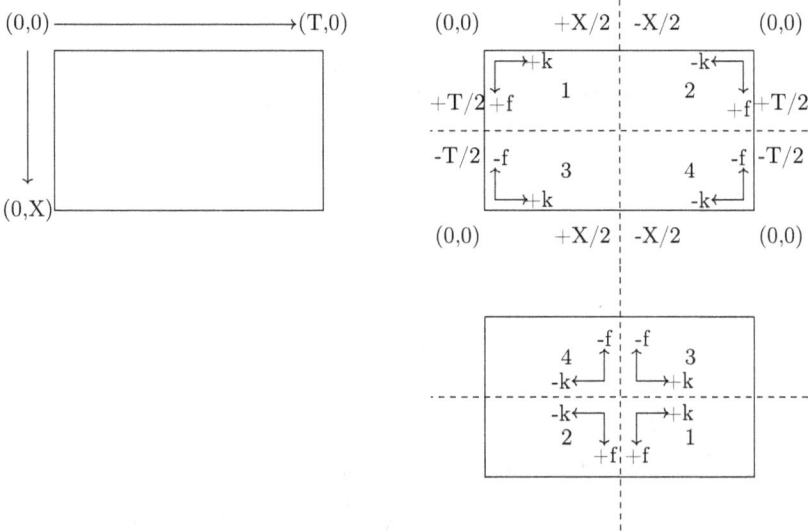

Figure 5.2: Sketch showing the reorganisation of the quadrants for 2D FFT.

Y=fft2 (y)

which returns the two-dimensional Fourier transform of a matrix using the fast Fourier transform algorithm (5.20). As in the one-dimensional spectrum the two half are symmetric, when doing a two-dimensional transform the data can be divided into four quadrants (as shown in the sketch in figure 5.2 at the top right corner), which will be complex conjugate parts.

In order to have the zeros of the frequencies and the wavenumbers positioned in the middle of the plot, one has to reorganise the quadrants. The operation consists in swapping the quadrants diagonally so that 1 will swap with 4 and 2 with 3. The resulting plot can be seen in figure 5.2 in the bottom right corner. From this, we only want to keep the positive frequencies and both positive and negative wavenumbers which show the eastward or westward propagation of the motion.

5.2.3.2 Energy spectra-Helmholtz decomposition

This method, developed by Bühler et al. (2014), allows to decompose one-dimensional spectra of the horizontal velocities and buoyancy into a wave component, which consists of inertia-gravity waves, and a vortex component, which consists of horizontal flow in geostrophic balance.

The advantage of using this method is that it provides a separation between balanced and unbalanced motions by only using two of the velocity components, plus temperature measurements. Usually decomposition methods need the whole three dimensional field, which is not available from our experimental measurements and often is not possible to obtain by observations.

A Helmholtz decomposition of the meridional and zonal velocities can partition the flow into a rotational and divergent part. In our case, we can transform the cartesian coordinates (xy-plane) into polar coordinates (θr-plane), so we have the azimuthal velocity component u and the radial component v (which corresponds to the along-track and across-track velocities defined in the paper by Bühler et al. (2014)). These two velocities can be expressed in terms of the stream function ψ and the velocity potential ϕ

$$u = -\psi_y + \phi_x \quad \text{and} \quad v = \psi_x + \phi_y, \tag{5.21}$$

which is the general decomposition of a two-dimensional flow into its rotational and divergent components.

By working in the Fourier transformed space, the same can be done on the one-dimensional spectra and the result (see Bühler et al. (2014) for the complete derivation) is

$$\hat{C}^u(k) = \hat{C}^u_\psi(k) + \hat{C}^u_\phi(k) \tag{5.22a}$$

$$\hat{C}^v(k) = \hat{C}^v_\psi(k) + \hat{C}^v_\phi(k) \tag{5.22b}$$

where the two terms on the left-hand side $\hat{C}^u(k)$ and $\hat{C}^v(k)$ are the power spectra of the velocity components, and the terms on the right-hand side give, analogously, the rotational ($2K^\psi(k) = \hat{C}^u_\psi(k) + \hat{C}^v_\psi(k)$) and divergent ($2K^\phi(k) = \hat{C}^u_\phi(k) + \hat{C}^v_\phi(k)$) component of the spectra.

The power spectrum for each velocity component can be calculated, according to the Parseval theorem, as the square of the absolute value of the Fourier transform. In MATLAB we use the following code for a unitary sampling frequency

```
L=length(u);
NFFT=2^nextpow2(L);
yy=fft(detrend(u,NFFT));
Cu=2*abs(yy(1:NFFT/2+1).^2);
```

where only half of the frequencies (negative or positive) is considered, as the signal is real valued, and the total power is conserved by multiplying the frequencies by a factor 2.

The left-hand side of this system of equations (5.22) is know from the data, what we want to calculate are now the four unknown variables on the right-hand side. To do so, we need a set of two more equations. These can be found thanks to the incompressibility assumption of the flow which provides a link between the meridional and radial spectra of the flow. The additional set of equations is

$$\hat{C}^u(k) = \hat{C}^u_\psi(k) - k\frac{d}{dk}\hat{C}^v_\phi(k) \tag{5.23a}$$

$$\hat{C}^v(k) = \hat{C}^v_\phi(k) - k\frac{d}{dk}\hat{C}^u_\psi(k) \tag{5.23b}$$

The solutions in terms of the observed velocity spectra are

$$K^\psi(k) = \frac{\hat{C}^v(k)}{2} + \frac{1}{2k}\int_k^\infty \left(\hat{C}^v(\overline{k}) - \hat{C}^u(\overline{k})\right) d\overline{k}, \tag{5.24a}$$

$$K^\phi(k) = \frac{\hat{C}^u(k)}{2} - \frac{1}{2k}\int_k^\infty \left(\hat{C}^v(\overline{k}) - \hat{C}^u(\overline{k})\right) d\overline{k}. \tag{5.24b}$$

Since we have a discrete data set, the integrals are calculated numerically via the trapezoidal method implemented in the MATLAB function

Q=**trapz** (Y) ,

where the integration area is broken down into trapezoids spaced by the distance between each sampled point.

The next step is to decompose the flow into a linear wave and vortex part. For this, we need to calculate the total energy spectrum. We already have the kinetic energy. Since we measure temperature together with the horizontal velocity components, we are able to determine the corresponding potential energy spectrum, which reads

$$\hat{C}^b(k) = |\hat{b}(k)|^2/N^2, \tag{5.25}$$

where b is the buoyancy, g is gravity, and N the averaged time and space buoyancy.

By further assuming that the field is composed of plane waves, from which vertical homogeneity follows, we can use the dispersion relation of gravity waves (2.37) to derive the equipartition relation for the spectra as

$$\hat{C}^b(k) + 2K^\psi(k) = 2K^\phi(k) + \hat{C}_W^w(k) \approx E(k), \tag{5.26}$$

where $\hat{C}_W^w(k)$ is the vertical kinetic energy, which we will assume negligible. With this additional assumption, the hydrostatic wave energy spectrum is $E_W(k) \approx 2K^\phi(k)$.

The application of this decomposition method to the atmosphere-like experiment data is presented in section 7.2.7.2.

Chapter 6

Baroclinic, Kelvin and inertia-gravity waves in the barostrat instability experiment

Figure 6.1: Three layers in the barostrat experiment

In this chapter, the dynamics developing in the barostrat experiment (see description of the set-up in section 3.2.1) is investigated. The first part deals with the the large-scale wave regimes, particularly focusing on the different waves developing in the two baroclinically unstable layers at the top and bottom of the water column, and in the stably stratified region between them. The interactions among waves occurring at different fluid heights in the tank are also addressed. The second part, instead, focuses on inertia-gravity waves observed along the jet of the baroclinic waves.

The study presented here follows a series of experiments performed at the BTU laboratories the results from which are published in the paper by Vincze et al. (2016) (from now on referred to as barostrat 1). In this first investigation, the combined effect of double convective and baroclinic instabilities in a rotating stratified layer are studied for the first time in a laboratory experiment. A series of questions arose after the barostrat 1. More specifically: two convective cells have been found at the top and the bottom of the tank, separated by a motionless layer. However, the cell at the bottom appeared to be baroclinically stable and the investigation on the differences between the top and bottom layers where not conclusive. Furthermore, a qualitative comparison between PIV measurements of the velocity field with the numerical simulations of the baroclinic instability in the atmosphere by O'Sullivan and Dunkerton (1995) showed similarities with the spontaneously emitted gravity waves. However, further analysis is needed in order to prove that the observed small-scale waves are indeed IGWs and that the barostrat instability can generically generate IGWs.

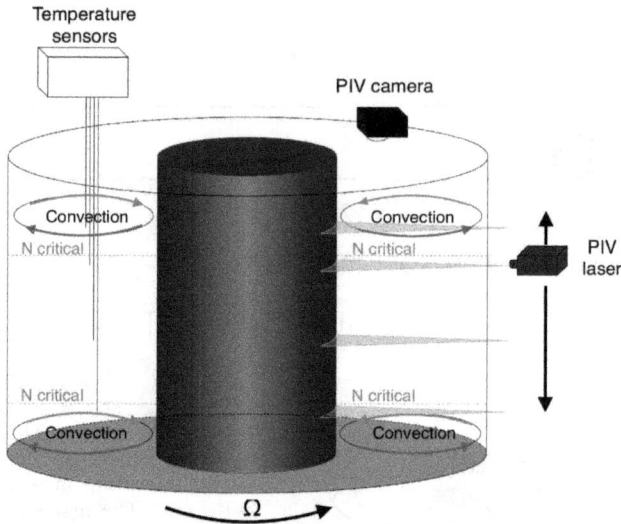

Figure 6.2: Sketch of the double-diffusive convection in the barostrat experiment and
of the experimental setup with PIV levels and temperature sensors positions. The red
dashed lines indicate the interfaces between the convective regions and the calm region
at the heights ztop = 8.9 cm and zbottom = 2 cm

The following chapter report the investigation done to address these open questions
and fill the gaps.

The main results presented in this sections are part of the paper *Baroclinic, Kelvin
and inertia-gravity waves in the barostrat instability experiment* by C. Rodda, I.D. Borcia,
P. Le Gal, M. Vincze and U.Harlander, published in *Geophysical & Astrophysical Fluid
Dynamics* (Rodda et al. 2018).

6.1   Set-up and measurements

The experimental apparatus has been already illustrated in details in section 3.2.1 where
how and why convective cells form at the top and the bottom of the water column forming
three distinct dynamic layers (visible in figure 6.1) are also explained. In this section we
will show some details regarding the measurement instruments and their positioning on
the experiment.

Both temperature sensors (see section 4.1.2) and PIV measurement techniques (see
section 4.2) are used to investigate the flow regimes at different water depths. For the
PIV, the green laser (Linenlaser LLM-SET-50-532, see technical specifications in the first
column in table 4.3) is fixed on a co-rotating vertical bar mounted at the outer cylinder
and can be moved at the desired fluid height during the experiment by remote control. The
laser produces a continuous horizontal light plane of 1 mm thickness which illuminates the

tracer particles (a mixture of hollow glass and silver coated hollow glass spheres, see details in table 4.4) in the fluid. The camera (GoPro Hero 4, screen resolution 1920 × 1080, fps = 30, without additional lens) is co-rotating and fixed at close range to the water surface. The field of view recorded by the camera covers approximatively one third of the annulus and allows to gain close-ups of the velocity fields, giving a better resolution in particular on the small-scale features. The recorded videos, having each a total duration of 12 minutes, are processed with the free Matlab toolbox UVmat (see section 4.2.2) and the horizontal components of the velocities are obtained. The velocities are measured at four fluid heights: 94 mm, 75 mm, 47 mm and 21 mm from the bottom of the tank. A sketch of the PIV system can be seen on the right hand side of figure 6.2.

The PIV particles are added to the fluid when the density profile is prepared at the beginning of the experiment. Considering that the experiment runs for maximum 6 hours, at least the smallest particles are expected to be found 1.5 cm below the surface and above the bottom of the tank at the end of it. In addition to the settling velocity, we have to take into account that the convective motions at the top and the bottom of the tank are mixing the PIV particles rather effectively. Therefore, at the measurement heights the particles remain in neutral buoyancy during the measurement time $\simeq$ 12 minutes.

The error on the PIV given by UVmat (see section 4.2.2.1) for the barostrat data is $rms = 0.5 - 0.6$ pixels for a typical displacement of 5 pixels. This gives an estimation of the PIV error $10\% - 12\%$. The percentile of excluded vectors is less than 1.5%. By using the a-posteriori error estimation, one obtain an error that is less than 15% over all the domain, fully in agreement with the UVmat output.

Four temperature sensors ALMEMO® are placed at the same fluid heights of the laser in the centre of the annulus gap, and diametrically opposite to the side where we do the PIV measurements. Hence the sensors, having a diameter of 500 µm and a Reynolds number of the order $Re = 1$, do not affect the PIV measurements. The temperature sensors have a sampling time interval $\Delta t = 1$ s and are able to record the data for the entire duration of the experiment (about 2 h). Thanks to the temperature sensors, we have long time measurements simultaneously at all the chosen fluid heights. A schematic drawing of position of the temperature sensors is given on the left hand side in figure 6.2.

6.2 Propagating waves at different fluid heights

This section presents the investigation of the flow regimes developing in the three different layers developing in the barostrat experiment. The reason for this analysis is the puzzling result found by Vincze et al. (2016) who observed baroclinic waves only in the upper convective layer (see figure 3.11), but not in the bottom layer. One possible reason for this is the damping due to bottom Ekman layer effects, however no further investigation regarding this matter had been done in this first series of experiments. In order to obtain a clear picture of the waves developing at the different fluid heights in the tank, we first consider frequency spectra of the horizontal velocity, obtained as described in Section 5.1.1, from which we can identify the dominant frequencies at each height (i.e. measured from the bottom of the tank: 94 mm, 75 mm, 47 mm and 21 mm).

We have separated the spectra into two windows: $0 < \omega < f$ (figure 6.3 left) and $f < \omega$ (figure 6.3 right). The grey dashed lines in the two middle plots on the right-hand side indicate the value of the buoyancy frequency N at the measurements heights. Peaks for low frequencies appear in the upper layer, for the frequency $\omega = 0.03\Omega$, at heights

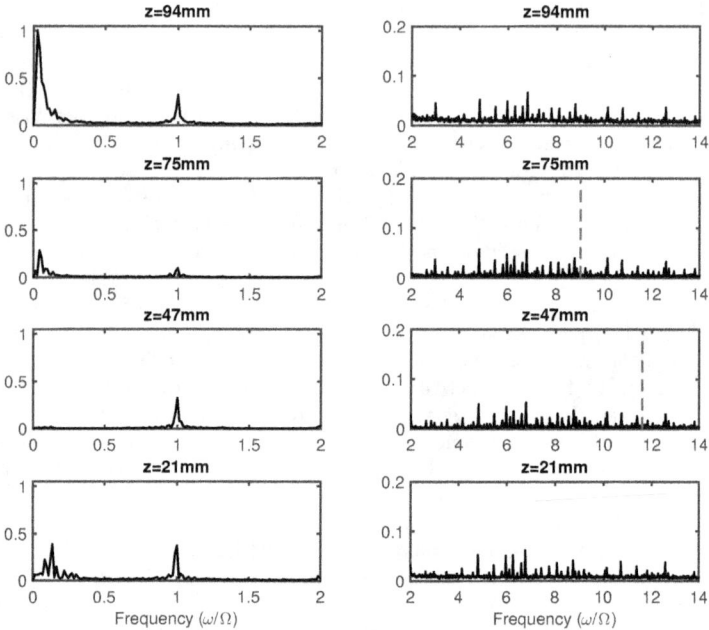

Figure 6.3: Velocity frequency spectra for the different heights. On the left the low frequencies, $0 < \omega < f$, on the right heigh frequencies, $f < \omega$. Note that the spectra amplitude on the left plot differs from the one on the right. The vertical dashed line indicates the value of N.

$z = 94$ mm (stronger) and $z = 75$ mm (weaker) and in the bottom layer ($z = 21$ mm), for the frequency $\omega = 0.14\Omega$. In the middle layer ($z = 47$ mm), instead, there is no peak in the low frequency range. Another peak, at $\omega = \Omega$, can be seen for all the layers, with a stronger signal at the bottom of the tank. Finally, in all layers there are peaks at frequencies larger than f.

Using the harmonic analysis as described in section 5.1.1, we can reconstruct the velocity fields corresponding to the most prominent peaks. The left column in figure 6.4 shows the baroclinic waves, one in the uppermost layer ($z = 94$ mm and $z = 75$ mm) close to the surface, and one in the bottommost layer, with different azimuthal wavenumbers, $m = 3$ and $m = 4$ respectively. Compared to an experimental run where the same rotation rate ($\Omega = 4$ rpm) and lateral temperature difference ($\Delta T = 10$ K) were applied, but we used pure water, so that one single convective cell develops covering the whole fluid depth, the baroclinic wave observed in the barostrat has a greater wavenumber. This observation is in full agreement with the correspondent Ro_T and Ta numbers calculated for both experiments using (3.17) and (3.16).

	Ro_T	Ta	wavenumber m
Pure water	2.16	1.57×10^7	2
Bottom layer barostrat	0.39	8.68×10^7	4
Top layer barostrat	0.31	1.11×10^8	3

Table 6.1: Comparison of the wavenumbers and Ro_T and Ta calculated with (3.17) and (3.16) for the experiment with pure water and the two baroclinically unstable layers at the top and the bottom of the barostrat experiment.

Indeed, from table 6.1, one can see that the Ro_T values for the two baroclinically unstable layers in the barostrat experiment are below the value for the pure water, and vice versa the Ta are above it. Amplitude vacillations, being precursor to a mode transition to a different wavenumber, develop through a supercritical bifurcation from the steady wave flow with the same wavenumber. For this reason they might appear at higher Ro_T and at lower Ta. Früh (2014) reported of an amplitude vacillating wave $m = 4$ for higher Ro_T and lower Ta with respect to a steady wave $m = 3$, in agreement with what we observe in our barostrat experiment.

The middle column of figure 6.4 shows the velocity pattern corresponding to the peak at frequency $\omega = \Omega$. We identify this pattern as inertial Kelvin waves modified by stratification. There are several examples in the literature of inertial Kelvin mode, also called the spin-over inertial mode, driven by precession either due to the rotation of the laboratory by the Earth (i.e. in a spherical shell (Triana et al. 2012)) or in a precessional cylinder (i.e. Lagrange et al. 2011). Moreover inertial Kelvin waves might be excited by elliptical instability, as reported by Lacaze et al. (2004). In our case the inertial Kelvin mode is more likely due to an imperfect alignment of the rotation axis with respect to gravity. In a system with a free surface this imperfection might force an inertial Kelvin mode even if a true precession of the rotation axis is absent (personal communication with Patrice Meunier).

Due to the fact that our measurements are not simultaneous, we do not have phase information and, hence, cannot investigate in detail the vertical structure of the inertial Kelvin wave and compare it with the analytical solutions found by Guimbard et al. (2010) for a rotating stratified cylinder. Nevertheless, our experiment shows that the amplitude of the inertial Kelvin waves varies with depth as can be noticed in figure 6.3 and figure 6.4 (see colorbar). This suggests that a structure of the wave along the vertical axis is indeed present. Moreover, it can be noticed in figure 6.4 that at the top ($z = 94$ mm and $z = 75$ mm) and at the bottom ($z = 21$ mm) of the tank the inertial Kelvin wave has a radial structure that differs from the middle region ($z = 47$ mm), namely it shows a higher radial wavenumber. This spatial modulation along the radius might originate from the existence in the top and bottom layers of the baroclinic instability. Gula et al. (2009) and Flór et al. (2011) reported instabilities resulting from resonances between Rossby and inertial Kelvin eigenmodes for a two layer flow in a rotating annulus. However, no clear signal of such resonance can be observed in our experiment, in particular since the mean Rossby number is too small and this resonance just happens for high Ro, although some interactions are detected.

In the right column of figure 6.4 the reconstructed velocity field for one of the main peaks with frequencies larger than f, for instance $\omega = 6.8\Omega$, is shown. Similar structures can be found for the other prominent peaks for frequencies $\omega > \Omega$ and are present in

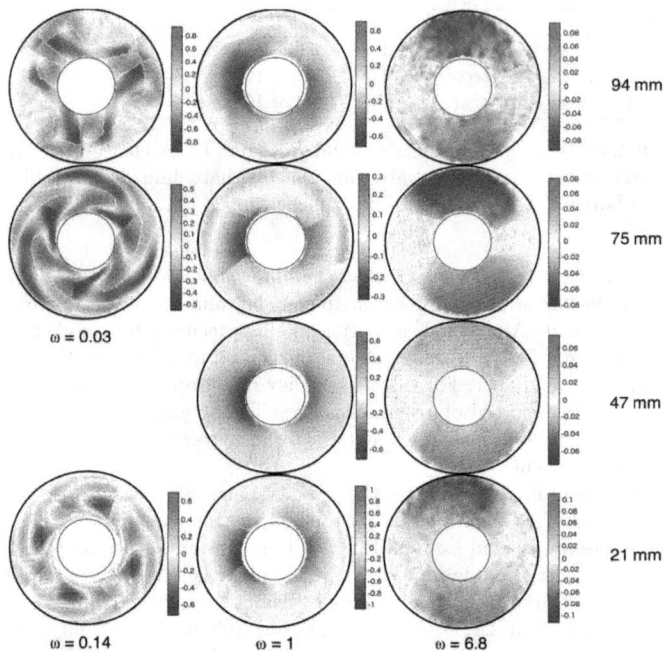

Figure 6.4: Reconstructed velocity field, using the harmonic analysis, for the principal peaks in the three layers. In the first column on the left the baroclinic waves, $m = 3$ in the uppermost layer (first and second lines from the top) and $m = 4$ for the bottom layer (bottom line in figure) are shown. The central columns shows the inertial Kelvin wave, for all layers. In the column on the right the field obtained for the main peak at high frequencies $\omega = 6.8\Omega$ is shown. All the frequencies are normalised by Ω.

all layers. It is also interesting to notice that most of these peaks appear at the same frequency, not changing with the height. Obviously, the horizontal structure is similar to the inertial Kelvin wave, but in this case the amplitude does not decrease in the direction of the outer wall and the wave propagates prograde. Remarkably, we found very similar high frequency waves in an experiment carried out with the same set up, but where we only rotate the cylindrical tank without any lateral temperature difference and vertical salinity stratification. We speculate that these waves with frequencies $\omega > f$ might be surface wave modes of the Poincaré type. They might be excited by a weak sloshing at the free surface and, in the case of the barostrat experiment, also at the interface between the layers of different density. We do not further investigate these weak gravity wave modes but will focus on the signature of *frontal* gravity waves that are very localised in space and time. Such localised wave packets move with the baroclinic jet and Fourier analysis of local time series is hence not a proper tool to detect those waves. More details on the IGW field related to the baroclinic front shall be given in section 6.3 after discussing the baroclinic wave dynamics in the next section.

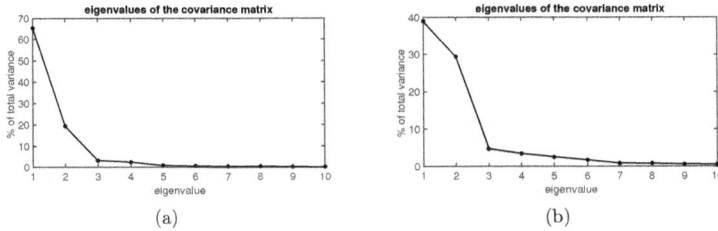

Figure 6.5: Variance of the first 10 eigenvalues of the coupled u and v velocities for the top layer (a) $z = 94$ mm and the bottom layer (b) $z = 21$ mm.

6.2.1 Determination of the baroclinic wave dynamics in the two unstable layers

In this section we investigate in more detail the similarities and differences between the baroclinic waves appearing in the top and bottom layer.

To do so, we performed an EOF analysis (see Section 5.1.2) of the horizontal velocity field measured in the uppermost layer in the tank (6.5a) and in the bottommost layer (6.5b).

Figure 6.5 shows the first ten eigenvalues of the covariance matrix. Each eigenvalue stands for the variance that can be explained by the corresponding complex function, the CEOF.

For the top layer ($z = 94$ mm), the first and the second eigenvalues, explaining together 85% of the total variance, are related to the baroclinic mode and its first harmonic (a power spectrum of the PCs shows one single peak for each PC at $\omega = 0.04\Omega$ and $\omega = 0.08\Omega$ respectively, these peaks correspond to the ones detected previously by the FFT analysis). The third eigenvalue, explaining the 3.4% of the variance, corresponds to the second harmonic of the baroclinic wave and the inertial Kelvin wave (two peaks in the PC spectrum at $\omega = 0.12\Omega$ and $\omega = \Omega$). The fourth eigenvalue, explaining 2.6% of the variability, is related to the inertial Kelvin mode.

The eigenvalue spectrum for the bottom layer ($z = 21$ mm) is quite different, as can be seen in figure 6.5b. Also in this case, the first eigenvalue is related to the dominant baroclinic mode, but explains only 39% of the total variance. The power spectrum of its correspondent PC shows a multitude of peaks with frequencies very close to the ones shown in figure 6.8(a), and in addition a peak for $\omega = \Omega$, suggesting a complex dynamics and interactions among waves. We shall discuss these interactions in more detail in the following text. The second eigenvalue, 29% of the variance, is related to the inertial Kelvin mode interacting with the baroclinic wave (the PC spectrum shows a peak for $\omega = 0.14\Omega$ and $\omega = \Omega$). The third and the fourth, explaining the 4.8% and 3.4% of the total variance respectively, show a broad spectrum of frequencies and are not easy to interpret as physical modes. The fifth and sixth eigenvalues, explaining 2.6% and 1.8% of the variability, are related to the second harmonic of the inertial Kelvin mode $\omega = 2\Omega$.

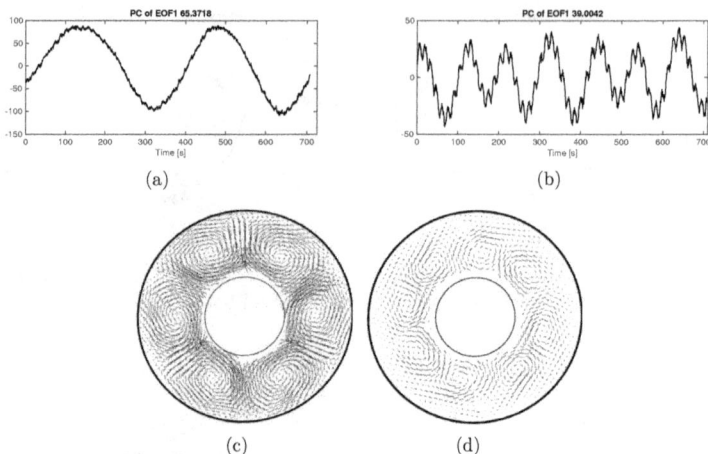

(a)

(b)

(c)

(d)

Figure 6.6: Comparison between the first EOFs and PCs for the top layer ($z = 94$ mm) and the bottom layer ($z = 21$ mm). (a) PC of the first eigenvalue $z = 94$ mm, (b) PC of the first eigenvalue $z = 21$ mm. (c) Real part of the EOF of the first eigenvalue at $z = 94$ mm and (d) real part of the EOF of the first eigenvalue at $z = 21$ mm

With respect to the eigenvalues of the covariant matrix for the uppermost layer (figure 6.6a) we have a significant reduction of the first eigenvalue. Moreover, an interaction between the baroclinic wave and the inertial Kelvin wave seems to play an important role in the dynamic of this layer, as both frequencies are contributing to the second CEOF.

A similar behaviour, for which the dominant component is smaller by about 20% compared to the steady wave regime has been observed by Hignett (1985) in case of an amplitude vacillating regime (AV). If we now consider the first CEOFs for the top layer (figure 6.6c) and for the bottom layer (figure 6.6d) and their respective PCs (figure 6.6a and 6.6b) we can see that in the top layer the fluid is in a steady wave regime, where the baroclinic wave shows a regular behaviour and has a very low phase speed (0.0057 rad/s), whereas in the bottom layer the baroclinic wave shows a variation of the amplitude in time with a phase speed of 0.0157 rad/s. Usually, a metastable transient AV has a very high phase speed, circa 5 times faster than the finally equilibrated flow (Früh and Read 1997). This supports our assumption that the flow in the bottom layer is in the vacillation regime.

For AV the amplitude of the wave varies periodically while the shape of the wave pattern remains constant. The strength of the vacillation is characterised by a vacillation index I_v defined over one vacillation cycle as

$$I_v = \frac{A_{max} - A_{min}}{A_{max} + A_{min}}. \tag{6.1}$$

Hignett (1985) indicates as critical vacillation index $I_v = 0.05$. This value separates flows in steady regimes (for $I_v < 0.05$) from flows in amplitude vacillation regimes (for

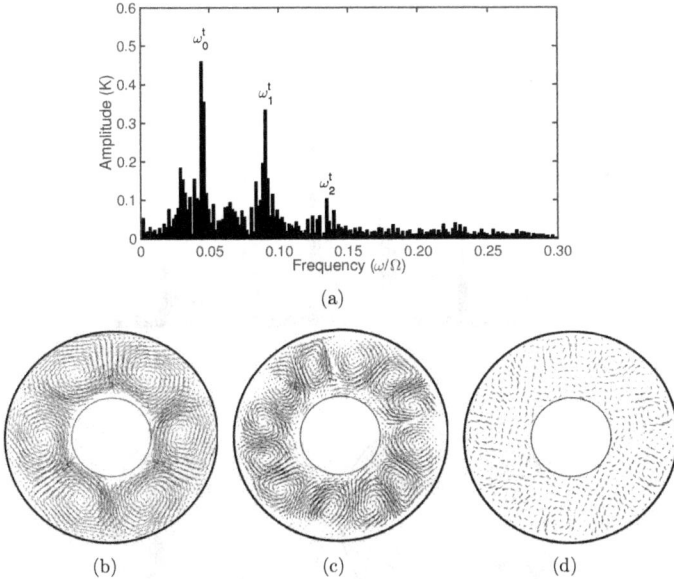

Figure 6.7: Low frequencies spectra from temperature data at $z = 94$ mm (a). The three peaks correspond to a baroclinic wave and its first and second harmonics: $\omega_0^t = 0.04\Omega$, $\omega_1^t = 0.08\Omega$ and $\omega_2^t = 0.12\Omega$; the velocity fields corresponding to these most energetic modes reconstructed using the CEOF analysis are plotted in (b),(c) and (d) respectively. The three waves of frequencies ω_0^t, ω_1^t and ω_2^t (at $z = 94$ mm) form a harmonic triad with azimuthal wavenumbers $m_0^t = 3$, $m_1^t = 6$ and $m_2^t = 9$.

$I_v > 0.05$). For the bottom layer of our experiment the value for the vacillation index is $I_v = 0.19$, while for the top layer $I_v = 0.03$.

For a steady wave regime it has been observed in many baroclinic wave experiments (e.g. Hide et al. 1977, Hignett 1985, Früh and Read 1997)) that the amplitude spectrum is composed almost entirely of the dominant component and its harmonics. This is also the case for the upper layer of our barostrat experiment. The peak corresponding to the dominant wavenumber $m = 3$ and its harmonics are dominating the low frequency spectrum plotted in figure 6.7(a) for temperature measurements at $z = 94$ mm. Figure 6.7 shows the reconstructed waves: baroclinic, first harmonic, and second harmonic with frequencies and wavenumbers $\omega_0^t = 0.04\Omega$, $m_0^t = 3$, $\omega_1^t = 0.08\Omega$, $m_1^t = 6$ and $\omega_2^t = 0.12\Omega$, $m_2^t = 9$ respectively. Differently from what was observed by Hignett (1985) and Hide et al. (1977), for which in most of the experiments the second harmonic was strongly pronounced, in our case the fundamental wave and the first harmonic are the most energetic ones. Moreover, it has been shown by Früh and Read (1997) that for steady waves the harmonic triad shows a more persistent and pronounced phase locking than the long wave triads.

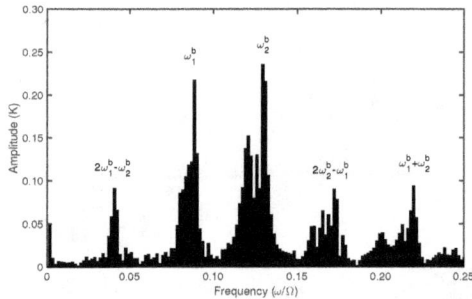

(a)

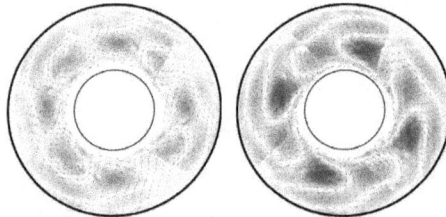

(b)   (c)

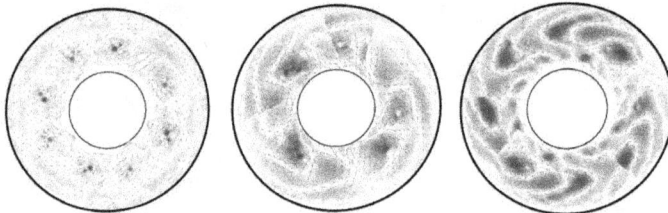

(d)   (e)   (f)

Figure 6.8: (a) low frequencies spectra from temperature data at $z = 21$ mm. The five peaks correspond to two baroclinic waves $m = 4$ (ω_1^b and ω_2^b), their nonlinear interactions with the mean flow (the two peaks $2\omega_1^b - \omega_2^b$ and $2\omega_2^b - \omega_1^b$) and the nonlinear interaction between the two main waves (at $\omega_1^b + \omega2^b$). The velocity fields corresponding to these peaks, reconstructed using the harmonic analysis, are shown in (b) $\omega_1^b = 0.089\Omega$, azimuthal wavenumber m=4, (c) $\omega_2^b = 0.13\Omega$, azimuthal wavenumber m=4. that are corresponding to the two baroclinic waves. (d) $2\omega_1^b - \omega_2^b = 0.048\Omega$, azimuthal wavenumber m=4 and (f) $2\omega_2^b - \omega_1^b = 0.17\Omega$, azimuthal wavenumber m=4 show the wave-mean flow interactions and (f) $\omega_1^b + \omega_2^b = 0.22\Omega$, azimuthal wavenumber m=8 the triadic resonance.

Buzyna et al. (1989) have represented a particular case of amplitude vacillation regime as due to an interference of two waves with the same azimuthal wavenumber and different phase speed. This regime is referred to as interference vacillation. The basis of this approach has been that a travelling, modulated wave can be represented as a linear superposition of two waves of the same azimuthal wavenumber, but with different phase speeds or frequencies. Figure 6.8(a) shows the spectrum obtained from the temperature sensor at $z = 21$ mm. We identify two identical shaped baroclinic waves with $m = 4$ at the frequencies $\omega_1^b = 0.089\Omega$ and $\omega_2^b = 0.13\Omega$, indicated in figure 6.8(a). Three other peaks with significant variance are recognisable: at $2\omega_1^b - \omega_2^b = 0.048\Omega$, at $2\omega_2^b - \omega_1^b = 0.17\Omega$ and at $\omega_1^b + \omega_2^b = 0.22\Omega$. The first two frequencies indicate the nonlinear interaction of the baroclinic waves ω_1^b and ω_2^b and the mean zonal flow as reported in detail in Buzyna et al. (1989). Figures 6.8(d) and 6.8(e) show the velocity fields, obtained with the harmonic analysis, for the frequencies $2\omega_1^b - \omega_2^b$ and $2\omega_2^b - \omega_1^b$. The spatial pattern of a baroclinic wave $m = 4$ can be seen in both figures, as one would expect from the wave-mean flow interaction.

A second nonlinear interaction can be identified in our experiment: the two baroclinic waves (ω_1^b and ω_2^b) interact nonlinearly forming a triad $\omega_1^b + \omega_2^b$ that might become resonant. The reconstructed velocity fields are shown in figures 6.8(b), 6.8(c) and 6.8(f). As we have seen, the interaction scenario is more complex then in the case of a steady wave. Moreover, in addition to the linear interaction between baroclinic waves, also the inertial Kelvin wave seems to interact with them, as already discussed. Früh and Read (1997) also observed a complex interaction scenario in amplitude vacillation regimes, and found that in this case long wave triads are usually observed.

It is interesting to notice that interference vacillation has been observed by Harlander et al. (2011) in the classical configuration of the thermally driven annulus. However, in their case there was no indication that the two waves were coupled through nonlinear interactions but they appeared to be a linear superposition of two modes of different zonal wavenumber drifting at different speeds. In our thermohaline version of the experiment, on the contrary, the amplitude vacillation results from linear interactions between two waves having the same wavenumber leading to a mean zonal flow and nonlinear interactions between the two waves and the mean zonal flow, more in agreement to the results from Buzyna et al. (1989).

6.3 Inertia-gravity waves

So far we focused on the large-scale modes, and their interactions. However, besides the already discussed baroclinic waves and inertial Kelvin waves, it is instructive to investigate IGWs occurrence at the baroclinic wave fronts. Such waves are very localised in space and time. In the paper by Vincze et al. (2016) wave trains whose characteristics are compatible with IGWs have been detected and a qualitative comparison between the PIV measurements with numerical simulations of spontaneous IGW emission from baroclinic fronts by O'Sullivan and Dunkerton (1995) show a qualitative good agreement.

6.3.1 Wave trains at $z = 94$ mm

We start the discussion on small-scale structures considering the upper most layer, at $z = 94$ mm, and in the next section we shall present the results obtained for the height

(a)

(b)

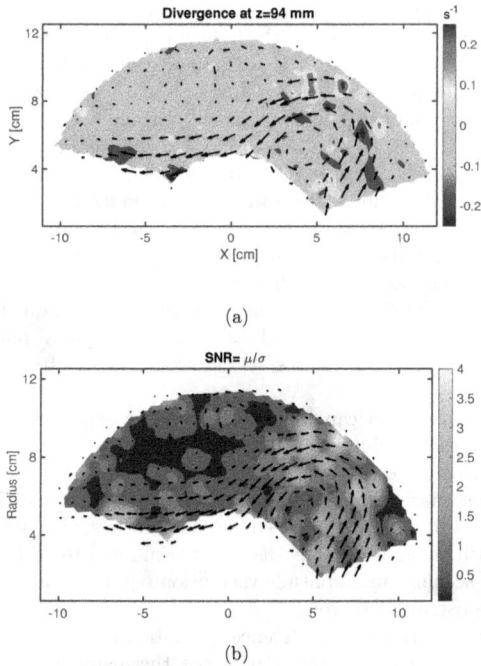

Figure 6.9: (a) Horizontal divergence at $z = 94$ mm, for $t = 535$ s. The arrows are showing the velocity field while the colour map represents the horizontal divergence. The horizontal wavelengths are: $(\lambda_x = 1\,\text{cm}, \lambda_y = 0.8\,\text{cm})$. (b) Signal-to-noise ratio, defined as $SNR = \mu/\sigma$. In the region along the jet the signal of small waves can be clearly distinguished from the background noise.

$z = 75$ mm. The other measurement heights are not presented because no clear signal of small-scale wave trains has been found.

In order to identify possible IGW signals we consider the divergence of the horizontal velocity field shown in figure 6.9(a) for the measurement height $z = 94$ mm. First of all, we show that the IGW signal detected in the horizontal divergence is above the noise threshold. To compare the level of the IGWs signal to the level of the background noise we use the signal-to-noise ratio defined as

$$SNR = \frac{\mu}{\sigma} \tag{6.2}$$

where σ is the square root of the mean variances of the background noise, i.e. calculated in an area where no IGWs signal is visible in the divergence field and μ is the locally (over a square running filter of dimensions 10×10 mm) averaged divergence field. The field of

SNR is plotted in figure 6.9(b), from which one can clearly see values above 1 (considered the threshold for which the signal can be distinguished from the background noise) in the region along the baroclinic jet, where the IGWs are found.

It can be noticed that for the shown layer most of the divergence signal is associated with the small-scale waves. Moreover, we can see a wave train structure clearly related to small-scale phenomena attached to the baroclinic jet. Similar structures are visible in most of our data, embedded in the baroclinic wave.

It is instructive to further investigate how these small-scale waves travel with respect to the baroclinic jet. The reason is that observations in the atmosphere by Uccellini and Koch (1987) identified intense low frequency IGW in the jet exit regions and hypothesised spontaneous imbalance as source mechanism. Other observations by Fritts and Nastrom (1992) have highlighted a conspicuous enhancement of gravity wave activity in the vicinity of jets and fronts and our experimental study seems to be in line with these numerical and observational findings. To examine these regions, at $z = 94$ mm, we took two different cuts, one parallel to the entrance of the baroclinic wave (figure 6.10) and one parallel to the exit of the baroclinic wave (figure 6.11). Because we record our data in the system of reference co-rotating with the tank and the baroclinic wave moves prograde, the wave front is crossing our line just for a quite short time, as it can be seen in figures 6.10(a), 6.10(b), and 6.11(a), 6.11(b). Hence, in the period 10 s $< t <$ 20 s in figure 6.10(c) and 15 s $< t <$ 30 s in figure 6.11(c), when the red line in figure 6.11(a) and 6.11(b) is along the jet, the most prominent travelling waves can be found. For a quantitative comparison, we plotted in figure 6.10(c) and 6.11(c) the drift speed of the baroclinic wave (dashed line) and the mean value of the zonal flow in the jet region (red line). Remarkably, the phase velocity of the small-scale waves (blue lines) are in both cases similar to the ones of the zonal flow. Compared to the baroclinic wave, the wave packets move faster, with phase speed equal to 3 mm s^{-1} at the entrance and 5 mm s^{-1} at the exit region of the baroclinic wave respectively, but are attached to it.

We can now apply a two dimensional Fast Fourier transform (see section 5.2.3 for more details) to the space time data in figure 6.10(c) and 6.11(c) to investigate the frequencies and the wavenumbers of the small-scale waves. The frequency - wavenumber plots are shown in figures 6.10(d) and 6.11(d) for the entrance and the exit region of the baroclinic wave respectively. To verify whether these observed waves satisfy the properties of IGWs we plot the dispersion curves taking into account the Doppler shift. According to this, we can now rewrite 1.1 as

$$\left(\omega_m - u_0 k_{\parallel}\right)^2 = \frac{N^2 k_{\parallel}^2 + f^2 n^2}{k_{\parallel}^2 + n^2}, \tag{6.3}$$

where ω_m is the frequency measured in the co-rotating system of reference, u_0 is the wind speed and $k_{\parallel}$ the wave vector along the magenta line in figures 6.10(a) and (b), and 6.11(a) and (b). Considering $N = 0.1$ rad s^{-1}, a maximum zonal flow in the jet $u_{\max} \simeq 3$ mm s^{-1} and minimum $u_{\min} \simeq 1$ mm s^{-1} consistent with the PIV measurement, the measured frequency $\omega_m = 1/6$ s$^{-1} = 1.04$ rad s^{-1}, the measured wavevector $k_{\parallel} \simeq 3$ cm^{-1}, and hypothetic vertical wavelength $\lambda_z^{94} = 2\pi/n = 5$ cm we obtain the dashed curves plotted in figures 6.10(d) and 6.11(d) in red and black, for $u_{\max}$ and $u_{\min}$, respectively. It can be noticed that in both cases the dispersion curves are following the behaviour of the signal showing a marked asymmetry between positive and negative wavenumbers. This asymmetry is due to the direction of propagation of the small-scale wave trains that is

towards the outer wall of the cylinder when situated at the entrance of the baroclinic jet
(figure 6.10) and in the opposite direction, i.e. towards the inner wall of the cylinder,
when situated at the exit of the baroclinic jet (figure 6.11). In both cases the strongest
peaks are are found within the two dispersion curves. More specifically, for the entrance
region some of the strongest peaks lie at lower frequencies closer to the red curve plotted
for $u_{min} = 1$ mm s^{-1}. With this lower value of the velocity the dispersion curve fits
better the lower frequencies. We speculate that these lower frequencies might be related
to small-scale waves propagating in regions of the jet where the flow velocity is lower.
From figure 6.10(c) 6.11(c) we realised that the wave packet travels with the jet. This
is further in agreement with our founding for the frequency, where $\omega_m = 1.04$ rad s^{-1} $\simeq$
$u_0 k = 0.94$ rad s^{-1} that suggests small-scale waves travelling mainly with the jet.

We have seen in section 2.6.1.3 that regions where the Rossby number is locally larger
than one are most favourable for spontaneous IGW emission. For our data we can calculate
the local Rossby number as

$$Ro_L = \frac{U}{fL},\tag{6.4}$$

where $U = \sqrt{u^2 + v^2}$ is the local shear, $f = 2\Omega$ is the Coriolis parameter and $L = 1/5(b-a)$
is the typical jet width. Once the local Rossby number is calculated, its spatial and
temporal correlation with the gravity wave signal can be investigated. Figure 6.12(b)
shows the contour plots of $Ro > 2$ superimposed to the horizontal divergence Hovmöller
plot obtained from data taken along the mid-gap arc shown in red in figure 6.12(a). What
can be seen from figure 6.12(b) is that in regions where $Ro > 2$ the gravity wave activity
is enhanced. Such regions are located along the jet, and our analysis suggests that it is
possible that the observed gravity waves are spontaneously emitted from the baroclinic
jet. A more extended analysis on the generation mechanisms is done in the next section
for the waves observed at $z = 75$ mm.

(a)   (b)

(c)

(d)

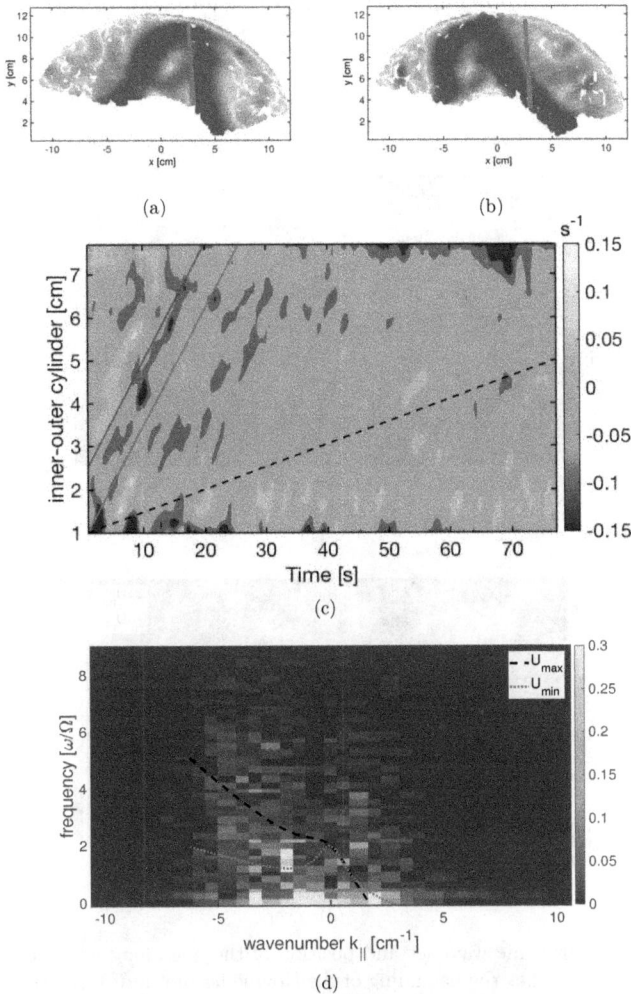

Figure 6.10: (a) Baroclinic wave jet and position of the line along which the Hovmöller plot is taken at $t_0 = 0$, i.e. the beginning of the Hovmöller plot in (c) and (b) at $t_1 = t_0 + 75$ s, i.e. the end of the Hovmöller plot in (c). Hovmöller plot for the divergence (in s^{-1}) at the entrance of the baroclinic wave at $z = 94$ mm. The dashed line is plotted to show the drift speed of the baroclinic wave, and the red line shows the mean value of the zonal flow in the jet region ($u_0 = 3$ mm s^{-1}). The phase velocity of the small-scale waves, measured from this plot and indicated by the blue line, is 3 mm s^{-1}, the same as u_0. (d) 2D spectra obtained from the space-time data of (c) plotted along with the dispersion curve (6.3) using $u_{max} = 3$ mm s^{-1} and $u_{max} = 1$ mm s^{-1} respectively. See text for more details.

(a)   (b)

(c)

(d)

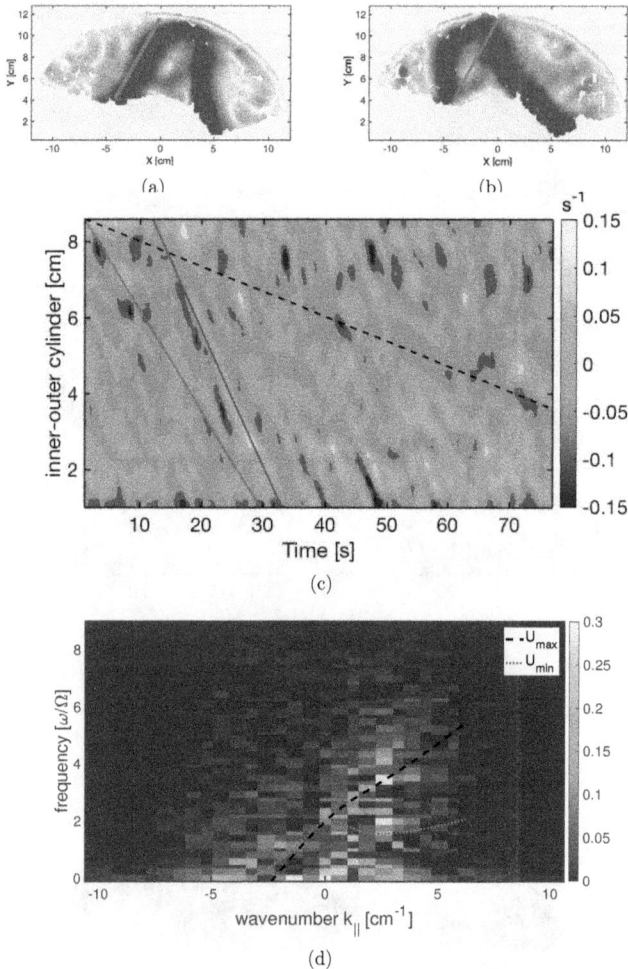

Figure 6.11: (a) Baroclinic wave jet and position of the line along which the Hovmöller plot is taken at $t_0 = 0$, i.e. the beginning of the Hovmöller plot in (c), (b) Baroclinic wave jet and position of the line along which the Hovmöller plot is taken at $t_1 = t_0 + 75$ s, i.e. the end of the Hovmöller plot in (c). Hovmöller plot for the divergence at the exit of the baroclinic wave at $z = 94$ mm. The dashed line is plotted to show the drift speed of the baroclinic wave, and the red line shows the mean value of the zonal flow in the jet region ($u_0 = 3$ mm s^{-1}). The phase velocity of the small-scale waves, measured from this plot and indicated by the blue line, is 5 mm s^{-1}. (d) 2D spectra obtained from the space-time data of (c) plotted along with the dispersion curve (6.3) using $u_{\max} = 3$ mm s^{-1} and $u_{\max} = 1$ mm s^{-1} respectively. See text for more details.

(a)

(b)

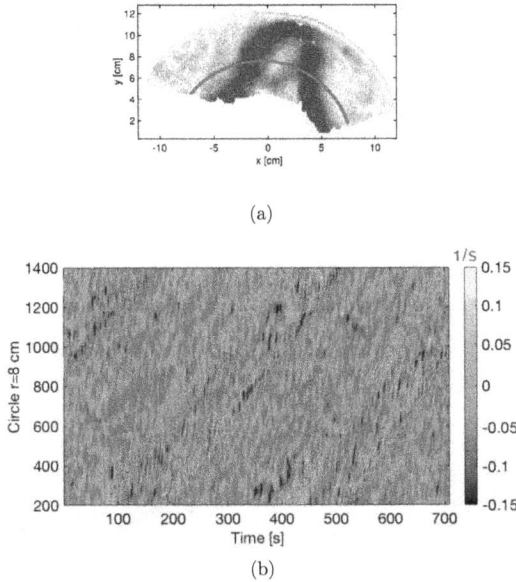

Figure 6.12: (a) circle of constant radius $r = 8$ cm along which the data are taken for the plot (b). (b) Hovmöller plot of the divergence at $z = 94$ mm, the red contour lines are $Ro > 2$. An enhanced IGW activity can be observed for regions where $Ro > 2$ indicating that spontaneous imbalance could be the generating mechanism responsible for such waves.

6.3.2   Wave trains at $z = 75$ mm

We continue the discussion on the small-scale waves signal in this section considering now the data at the measurement height $z = 75$ mm. We recall that this height is in the stratified region where $N > f$, for this reason IGWs are expected to show similarities with atmospheric gravity wave packets; the frequencies range for IGWs at this height is highlighted by the light blue coloured central region in figure 3.12(b). Furthermore we want to recall that the baroclinic wave with $m = 3$ is present at this heigh even though it is weaker and drifts in the opposite direction with respect to the baroclinic wave observed at the height $z = 94$ mm (see figure 6.4).

We follow the data analysis presented in the previous subsection discussing the analogies and the differences between the waves observed at the two heights.

Figure 6.13 shows the plot of the horizontal divergence (a) in a snapshot at the time $t = 706$ s where a wave train can be seen, similarly to the one found at $z = 94$ mm (see figure 6.9). The related signal to noise ratio, calculated with (6.2) as described in the previous section, is plotted in Figure 6.13(b). The maximum of SNR corresponds to the

(a)

(b)

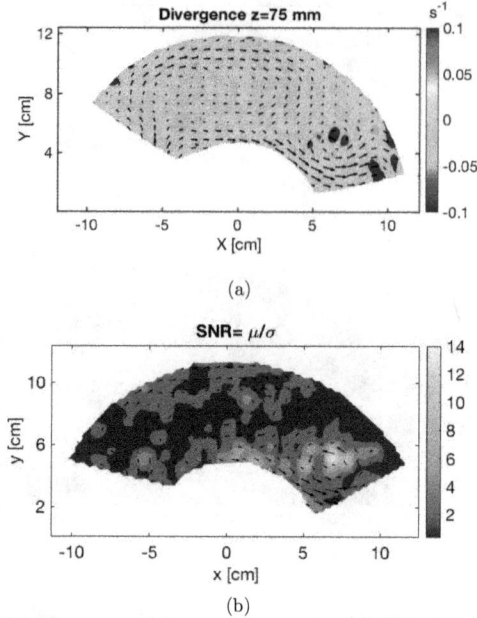

Figure 6.13: Horizontal divergence at $z = 75$ mm, for $t = 706$ s. The arrows are showing the velocity field while the colour map represents the horizontal divergence. The horizontal wavelengths are: $(\lambda_x = 1$ cm, $\lambda_y = 0.8$ cm). Signal-to-noise ratio, defined as $SNR = \mu/\sigma$. In the region along the jet the signal of small waves can be clearly distinguished from the background noise.

position of the wave train in the plot above, pointing to a clear distinction of the signal from the background noise. It can be noticed that, contrarily to what we observed for $z = 94$ mm, at this height the wave train is not positioned on the baroclinic wave jet, but in front of it.

Because the baroclinic jet is weaker at this fluid height and moreover the small-scale waves are not positioned along it we decided not repeat the analysis done for $z = 94$ mm where we have chosen two lines along the entrance and the exit regions of the jet (figure 6.10 and 6.11).

Instead, for the $z = 75$ mm level we chose an arc of the circle of radius $r = 8$ cm (see figure 6.14(a)), i.e. at the middle of the gap width, to construct a space-time diagram from which we derive a frequency-wavenumber diagram. The Hovmöller plot of the divergence along this arc is shown in figure 6.14(b) and the correspondent 2D spectra in figure 6.14(c). By Doppler shifting the dispersion relation using $u_0 = 1$ mm s^{-1} and $k = 5$ cm^{-1} we obtain the dispersion curves plotted with red, black and yellow dashed lines in figure 6.14(c) for vertical wavelengths $\lambda_z^{75} = 1, 0.3, 0.2$ cm respectively. Obviously, the curve with $\lambda_z^{75} = 0.2$ cm corresponds best with the IGW dispersion curve. We hence find that the estimated values for the vertical wavelengths at the two different fluid heights, $z = 94$ mm and $z = 75$ mm, are very different: $\lambda_z^{94} \simeq 5$ cm and $\lambda_z^{75} \simeq 0.2$ cm. This tells us that the upper layer waves have a horizontal phase speed and a vertical group speed (vertical particle motion). Since λ_z is large, and energy goes downward the waves can trigger motion in the lower layer. Here, due to the strong stratification, vertical motion is suppressed and the particles move horizontally (horizontal group velocity and vertical phase speed), implying a frequency close to f, a very small vertical wavelength and important dissipation. Therefore, the wave packets cannot move deep into the stratified layer. It is, hence, no surprise that we cannot see much wave activity in the stratified layers.

In figure 6.14(b) the red contour lines indicate the values of the local Rossby number exceeding the threshold $Ro > 1$. At this fluid height only two wave packets are visible at $t \simeq 400$ s and $t \simeq 700$ s, a different behaviour from the one observed in the upper most layer ($z = 94$ mm, figure 6.12) where the wave trains appeared to be emitted continuously from the jet stream where the Rossby number has typically values of 2. Despite this difference, in both cases a clear spatial-temporal correlation between the highest values of the local Rossby number and the waves emission is found, suggesting spontaneous imbalance as generating mechanism as stronger IGWs radiation is to be expected from regions of imbalance, where wind speeds are strong (O'Sullivan and Dunkerton 1995).

A complete analysis on the generation mechanisms of the observed short scale waves observed in our experiment is beyond the purpose of this paper. Nevertheless, some possible wave excitation mechanisms might be excluded. One is excitation due to convection since the gravity waves have not been found in our experiment setup without rotation, i.e. without the baroclinic jet. Kelvin-Helmholtz shear instability might also be responsible for wave emission. As reported by Williams et al. (2005), a good indicator of the Kelvin-Helmholtz shear instability is the dimensionless Richardson number:

$$Ri = \frac{N^2}{(du/dz)^2} \tag{6.5}$$

where $u(z)$ is the horizontal velocity profile. Even though we do not have enough measurements of the horizontal velocity along the water column (see figure 3.12(c)) and we do not have a precise value of N at $z = 94$ mm, we can use the values of the azimuthal velocity at $z = 94$ mm and $z = 75$ mm to calculate the vertical shear and the typical values obtained from the numerical simulation done by (Borchert et al. 2014) for $N/f = 0.1 - 0.3$ to estimate the Richardson number. For $z = 94$ mm, $Ri^{94} = 0.016 - 0.057$. This value, being smaller than 0.25, suggests an unstable stratified shear flow that would point to Kelvin-Helmholtz instability. However, the position of the small-scale waves with respect to the baroclinic jet and their occurrence related to an enhanced value of the local Ro number does not rule out spontaneous imbalance as generating mechanism. For $z = 75$ mm instead we have $Ri^{94} = 14$. Being this much larger than 0.25 rules out shear instabilities as generating mechanism and together with the large values of the local Rossby number correlated with the occurrence of the waves (figure 6.14(b)) strongly hints to an unbalanced dynamics.

Finally we want to mention that optical deformation of the free surface due to capillary waves would have a much larger phase speed than the waves we observed. We conclude that capillary waves are not responsible for the small-scale structures at the free surface.

(a)

(b)

(c)

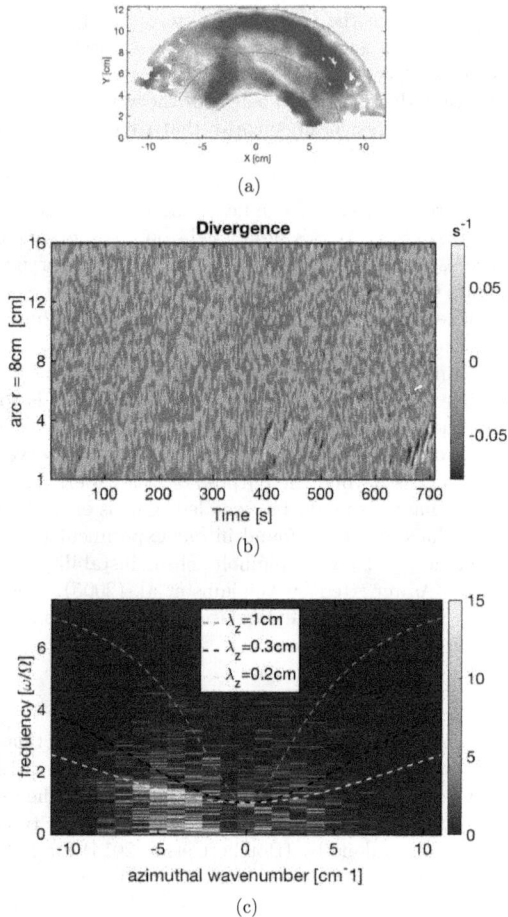

Figure 6.14: (a) circle of constant radius $r = 8$ cm along which the data are taken for the plots (b) and (c). (b) Hovmöller plot of the divergence at $z = 75$ mm, the red contour lines are $Ro > 1$. (c) 2D fft of the divergence space-time data plotted in (b). The dashed coloured lines show the dispersion relation for gravity waves (1.1) considering vertical wavelenghts $\lambda_z = 1$ cm, $\lambda_z = 0.3$ cm and $\lambda_z = 0.2$ cm.

Chapter 7

Atmosphere-like differentially heated annulus

This chapter is devolved to the work done with the big tank apparatus in its various set-ups already presented in section 3.3. We will start our analysis of the data collected with the original set-up and afterwards we will see the main differences/ analogies observed when the set-up is modified, particularly focussing on the baroclinic wave regimes and gravity wave signal.

7.1 Large-scale flow

The first experiments done once the experiment was built focussed on the investigation of the regime diagram, and in particular to study for which temperatures and rotation rates the transition between the axisymmetric regime and the wave regime could be observed. The data collected during the campaigns have been used for a comparison with data retrieved from the small-tank configuration, with other experiments in the literature, with theoretical predictions, and with numerical simulations.

The geometry and experimental conditions of the laboratory experiment are transferred to corresponding numerical simulations employing the numerical model *cylFloit* used by Borchert et al. (2014) and Hien et al. (2018). This model applies a finite volume algorithm to solve the Navier-Stokes equations for a Boussinesq fluid in an annular domain. The resolution of the regular grid in cylindrical coordinates is $N_\theta = 600$ azimuthal, $N_r = 200$ radial, and $N_z = 135$ vertical. The boundary conditions applied at the lateral and bottom walls are no-slip. The boundary condition at the top of the tank, simulating the free surface of the experiment, is the slip condition

$$\left.\frac{\partial u}{\partial z}\right|_{z=d} = \left.\frac{\partial v}{\partial z}\right|_{z=d} = 0, \quad \text{and} \quad w|_{z=d} = 0. \tag{7.1}$$

Therefore, the top condition in the numerical simulations consists of an 'inviscid' lid where the stresses tangential to the surface, surface waves, and the surface tension are neglected. Moreover, the top of the annulus is considered adiabatic; thus the heat transfer between the fluid and overhead ambient air is excluded from the model. The omission of wind stress and evaporative cooling might be a source of discrepancies between the laboratory experiment and the numerical simulations (as discussed later).

In order to obtain the best possible comparison between model and laboratory experiment, the simulation strategy is based on the experimental procedures. First, a 2D simulation is performed with a constant rotation rate of $\Omega = 0.1$ rpm until a steady state is established. Subsequently, a random disturbance is added to the temperature field before a 3D simulation of the entire annulus configuration is performed. The rotational speed is thereby linearly increased by 0.01 rpm every 15 s until the final speed is reached. Over the entire period, at each integration step, a random disturbance is applied to the inner and outer cylinder wall temperatures in order to mimic the temperature variations perceived in the experiment.

7.1.1   Regime diagram

To investigate dominant flow regimes, we run several experiments in the laboratory and numerically; the imposed temperature difference was kept constant, and we varied the rotation rate. For the imposed temperature difference, the smallest rotation rate for which a baroclinic wave sets in is $\Omega = 0.5$ rpm. A regular wave regime with azimuthal wave number varying between $4 \leq m \leq 7$, has been observed for rotation rates of $0.5 \leq \Omega \leq 1$ rpm (displayed in figure 7.1). For $\Omega > 1$ rpm, the flow starts to enter a geostrophic turbulent regime. By running the experiment several times under the same conditions, baroclinic waves with different wave numbers occur, suggesting the existence of multiple equilibria. The numerical simulations confirm these observations.

The minimum and maximum azimuthal wavenumbers observed are consistent with the geometry of the annulus, expressed by the empirical law (Hide and Mason 1970)

$$\frac{\pi}{4}\frac{b+a}{b-a} \leq m \leq \frac{3\pi}{4}\frac{b+a}{b-a}, \tag{7.2}$$

which gives $m_{\min} = 3$ and $m_{\max} = 7$ for our experiment. Since $m = 3$ has not been observed, we expect that the region for $m = 3$—for the chosen Burger number Bu and Taylor number Ta—is small. For other choices of rotation rate and lateral temperature difference, the region for $m = 3$ might be larger and therefore observed in the experiments, but we have not investigated this.

Figure 7.1 compares the regime diagram for the new atmosphere-like tank with the classical configuration (Vincze et al. 2015) and a rotating annulus experiment from (Scolan and Read 2017), where thermal forcing is applied by local heating and cooling sources. The theoretical instability curves from the O'Neil (1969) model are plotted for both configurations (grey line for the classic, with $m = 3$, and blue for the atmosphere-like rotating annulus, with $m = 4$). The straight red line corresponds to the inviscid instability criterium (3.3) by Eady for $m = 3$. It should be noted that the transitions shown in figure 7.1 are consistent with (O'Neil 1969) and that for both, the small- and big-tank, some baroclinic waves can be observed in the region of the diagram above the inviscid Eady line and below the O'Neil curves. Moreover, these theoretical curves show a very good agreement with the experimental curves from (Hide and Mason 1970) (plotted in black in figure 7.1). Note further that we observed an almost perfect correspondence between the experiments and the simulations (the markers labelled with $m = 0, 4/5, 6/7, 7$ in the regime diagram shown in figure 7.1 are for both the numerical simulations and the experiment) with respect to the azimuthal wave numbers.

Following Hide (1967), we define the dimensionless parameters $\sigma_z = \Delta_z T / \Delta T$ and $\sigma_r = \Delta_r T / \Delta T$, where $\Delta_r T = (b-a) \partial T / \partial r$ is the radial temperature gradient in the fluid interior multiplied by the dimension of the annular gap. Hide (1967) found heuristically that for high Péclet number, the parameters $\sigma_z \approx 0.67$ and $\sigma_r \approx 0.33$. These values

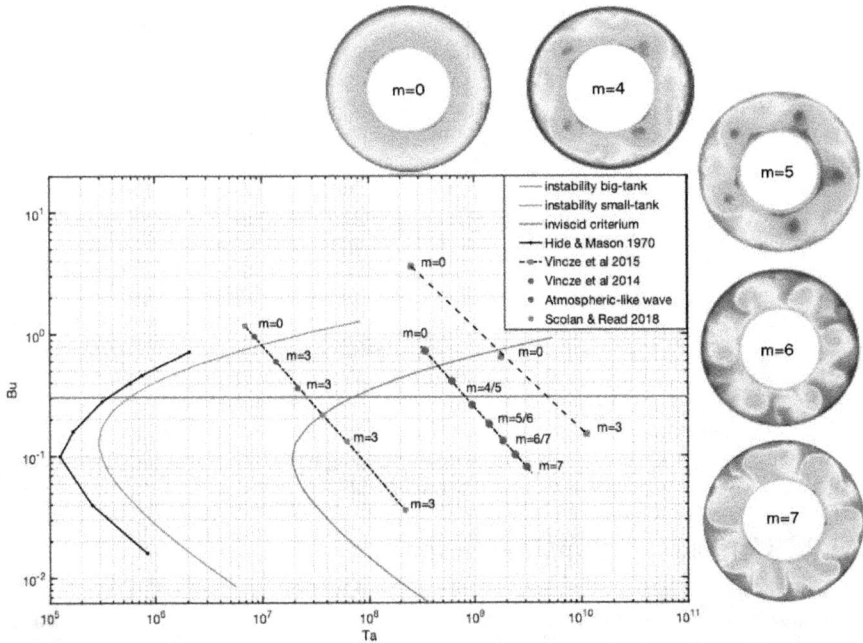

Figure 7.1: Regime diagram and instability curves. The straight red line is Eady's inviscid model from (3.3) for $l = 1, m = 3$, the grey and blue curves are from O'Neil (1969) for the small-tank and the big tank with constant $\epsilon_B = 0.3$ and $m = 3$ and $m = 4$ respectively. The Burger number of the experimental data for the small and big-tank is calculated assuming that $\Delta_z T = \sigma_z \Delta T$ constant for the different rotation rates, to have a better comparison with the model. The small-tank data are taken from Vincze et al. (2015) and Vincze et al. (2014) (darker red dots). The black curve is from experimental data by Hide and Mason (1970). Data from Scolan and Read (2017) are also plotted (green dots) for a comparison with a configuration when Π is < 1. Surface temperature data acquired with an infrared camera show wave flow regimes for different azimuthal wavenumbers found in the new big-tank configuration.

compare well with several laboratory experiments reported in their paper, which are however all limited to inverse aspect ratio $\Pi > 1$, i.e. tall annuli with narrow gaps and high fluid depths. Experimental studies for $\Pi < 1$ were done by Douglas and Mason (1973) for a differentially heated annulus with rigid upper lid, and they measured a consistent decrease of σ_z and increase of σ_r for decreasing fluid depths. More specifically, for $0.13 < \Pi < 0.2$, which is the range into which our experiment with $\Pi = 0.17$ is included, they measured $0.3 < \sigma_z < 0.5$. We can verify whether this value matches the temperature gradients measured in the atmosphere-like laboratory experiment. First of all, we calculate the Péclet number $Pe = g\alpha\Delta T\nu^{1/2}/8\kappa\Omega^{3/2} = 570$. Since Pe is used to distinguish regimes for which the flow is dominated by thermal conduction (low Pe) to the ones dominated by convection (high Pe), we can see that our experiment falls in the second type of flow. Then, we can calculate σ_z and σ_r and compare them with the values found by Hide (1967) and Douglas and Mason (1973). For an externally imposed $\Delta T = 4.8\,°C$ and $\Omega = 0.7$ rpm, the measured $\Delta_z T$ is in the range $1.7\,°C$ (close to the surface) to $1.3\,°C$ (close to the bottom) giving $\sigma_z = 0.27 - 0.35$. If we, instead, consider the horizontal difference of the vertically averaged temperature measured in the fluid interior (see $\overline{T}_w - \overline{T}_c$ in table 7.1), the value of the factor raises to $\sigma_r = 0.52 - 0.68$. Our results show excellent agreement with the investigation of Douglas and Mason (1973), confirming a decrease of σ_z and an increase of σ_r for shallow water configurations. Scolan and Read (2017) measured the stratification through the vertical temperature profiles at mid-radius and used those to calculate the Burger number plotted in figure 7.1 (green dots). It can be seen that the dashed line connecting the data is not parallel to the other two. This shows that the coefficient σ_z is not constant in the parameter regime, but it increases moving towards the bottom right corner in the $Ta - \beta$ diagram.

7.1.2   Comparison temperature and N

We now compare results from numerical simulations with the experimental data to investigate how well the model developed by Borchert et al. (2014) can reproduce the data in spite of deviations in the upper boundary condition and the omission of surface heat loss due to radiation and evaporation and the neglect of surface waves and surface tension. As we will see, the deviations are not too significant, and the qualitative structure matches with the experiment. Therefore, we can later use the numerical data to (at least qualitatively) understand better certain features we observe in the experimental data, e.g., the occurrence of gravity waves trapped along the jet axis (Rodda et al. 2018).

For this comparison, we select a laboratory experiment and the corresponding numerical simulation and investigate the temperature and N field more into detail.

For the numerical simulations, the boundary temperature at the walls is constant along the vertical and set to $T_b = 25.6\,°C$ at the outer wall and $T_a = 22.8\,°C$ at the inner wall; a random noise of amplitude $0.5\,°C$ is added to both. For the laboratory experiment the temperature measured by the sensors attached to the walls at two fluid depths $z = 5\,cm$ and $z = 1\,cm$ are $T_{w,z=5} = 24\,°C$ and $T_{w,z=1} = 21\,°C$ at the warm wall, $T_{c,z=5} = 20.2\,°C$ and $T_{c,z=1} = 19.8\,°C$ at the cold wall. From these measurements one can see that the lateral temperature difference closer to the free surface $T_{w,z=5} - T_{c,z=5} = 3.7$ is higher than the one close to the bottom of the tank $T_{w,z=1} - T_{c,z=1} = 1.4$. Although the numerical simulations do not capture this depth variation, the mean temperature values close to the boundary lateral walls compare reasonably well, as it can be seen in the second column in table 7.1.

It is worth mentioning that with N_v from table 7.1 and $H = 6\,cm$ we find $N^2 H/g \approx$

(a) experiment   (b) simulations noise

(c) experiment   (d) simulations

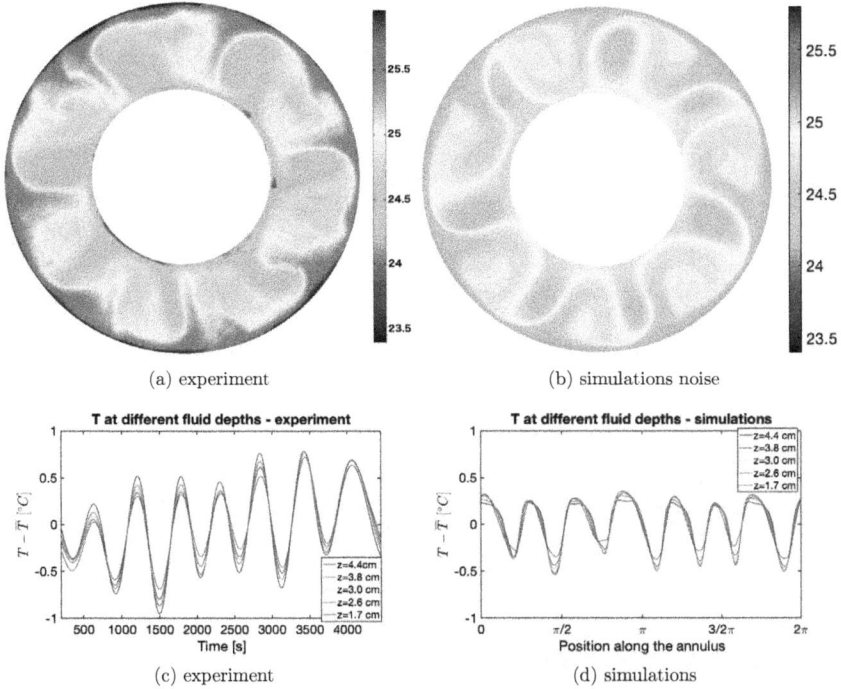

Figure 7.2: Surface temperature for the experiment (a) measured with the infrared camera and the numerical simulations (b). Temperature series normalised by the mean temperature ($\overline{T}_{\text{laboratory}} = 21.3\,^\circ\text{C}$ and $\overline{T}_{\text{simulations}} = 24.2\,^\circ\text{C}$) taken at five fluid depth along a vertical line at mid gap for the laboratory experiment (c) and simulations (d).

2.4×10^{-4} (for the ocean it is roughly 3×10^{-2}). It is known that for $N^2 H/g \ll 1$ the internal gravity wave spectrum is separated from the surface wave spectrum. Hence, the surface gravity waves, neglected in the numerical model, will not affect the IGWs (Pedlosky 2013).

The large-scale temperature patterns at $z = 6\,\text{cm}$ observed in the laboratory experiment and numerical simulation are shown in figure 7.2(a) and (b). The laboratory experiment temperature is measured with the infrared camera, and the temperature field for the entire annulus is reconstructed using subsequently recorded images. A leading baroclinic wavenumber $m = 7$ is found in both experiment and simulation, after a steady state is reached for the rotation rate $\Omega = 0.7$ rpm. The surface temperature compares qualitatively in view of the the fact that, in contrast to the experiment, the numerical model has a no flux boundary condition at the upper surface.

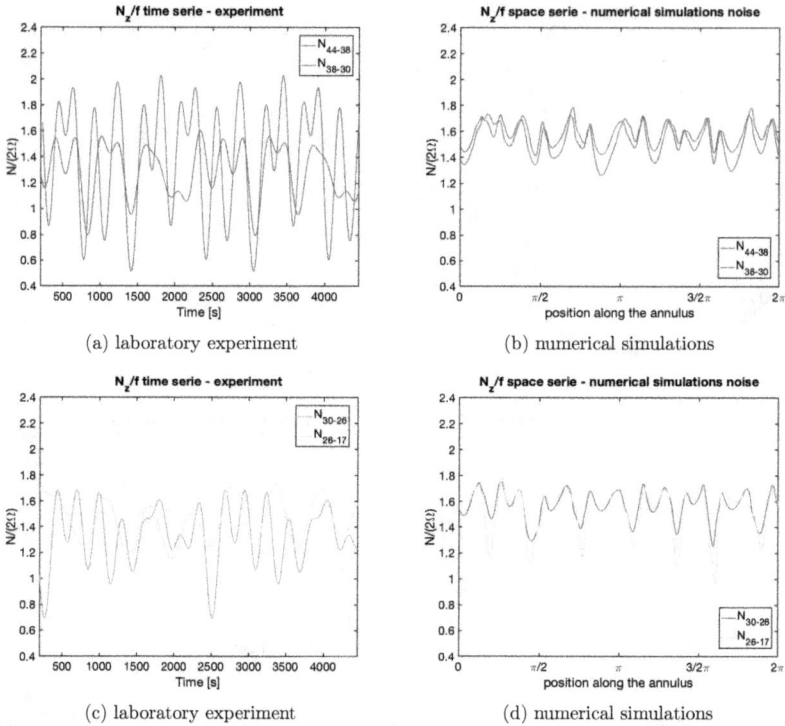

Figure 7.3: N/f calculated from the temperature for the upper most two measurements: experiment (a) and simulation (b); and the two bottom measurements: experiment (c) and simulation (d).

The next analysis presented is done by using the temperature data (where the mean value has been subtracted to each of them) taken from five fluid depth points along a vertical line at mid-gap width. The temperature time series for the period that the baroclinic wave needs to travel over the full azimuth from 0 to 2π is shown for the experiment in figure 7.2(c). This is compared with numerical simulations (figure 7.2 (d)), for which instead of measuring time series at a fixed point, we have taken the data along the entire circumference at mid-gap width at one fixed time. The comparison is made assuming that there is a spatio-temporal equivalence for a steady wave regime. The experimental data show larger amplitude variation with respect to the numerical simulations at all fluid depths. Nevertheless, the vertical temperature differences for the entire water column, calculated as $\Delta_z T = (T_{z=4.4} - T_{z=3.8})/\Delta z$ (see third column table 7.1), are consistent. From the measured temperature differences along the radial and vertical directions, we can calculate the buoyancy frequency. $N_H = [(g\alpha\Delta_r T)/H)]^{(1/2)}$ is the buoyancy frequency calculated using the radial difference of temperature measured outside the boundary layers at the walls ($\Delta_r T = \overline{T}_w - \overline{T}_c$, with $\overline{T}$ being the vertical mean between the two measurement points), whilst $N = [(g\alpha\Delta_z T)/H]^{(1/2)}$ is the buoyancy frequency calculated by using the

	f [rad/s]	$\overline{T}_w - \overline{T}_c$[°C]	$\Delta_z T$[°C]	N_H [rad/s]	N_v [rad/s]	m
Laboratory experiment	0.15	2.5	1.6 -1.5	0.29	0.21 - 0.22	7
Numerical simulations	0.15	2.8	1.5	0.31	0.22	7

Table 7.1: Comparison laboratory experiment (BTU) and numerical simulations (GUF).

vertical temperature difference $\Delta_z T$. This gives $N < N_H$ consistent with the previous observation that $\sigma_z < \sigma_r$. Another interesting feature of N is shown in figure 7.3, where on the left-hand side there are the plots for the laboratory experiment (a) for the two uppermost and (b) for the two bottom most measurements. On the right-hand side are the corresponding plots for the numerical simulations (b) and (d). The mean value for N is the same and it is greater than the Coriolis frequency f. However, the amplitude of the vacillations measured in the experiment is two times that of the oscillations measured in the numerical simulations, particularly when the data close to the water surface are considered (figure 7.3(a) and (b)). It is likely that these differences at the surface come from the differences of the upper boundary conditions found between simulation and experiment. Note that taking the values from table 7.1 we see that $N_H/\Omega \approx 4$. This value corresponds rather well with our estimate from figure 3.6(a) where we found, for $\Omega = 0.075$, $N/\Omega = 4.5$. The measured value for the vertical temperature gives $N/\Omega \approx 3$, and by taking this, the circles in figure 3.6(b) for the atmosphere-like tank move towards the value $\delta = 0.1$, indicating that we are close to the critical ratio between the Ekman layer and the total fluid depth with $d = 6$ cm, as used in our experiment. Indeed, for experimental runs with a total fluid depth $d = 4$ cm, no stable baroclinic waves could be observed. This shows that our conclusion taken in section 3.2 on the basis of simple scaling arguments, saying that only a tank with a small aspect ratio H/L provides a $N/f > 1$, is correct. Together with the experimental finding, namely that $H \geq 3$ cm for steady baroclinic waves in the small tank, we have experimentally shown that only a tank with 4 cm $< H << L$ is suitable to study atmospheric-like gravity wave emission from baroclinic jets.

7.1.3   Spatial structure of N/f and gravity wave trapping

To better understand the spatial variation of N/f in the annular domain, we plotted horizontal maps in figure 7.4 using the data from the numerical simulation. A similar pattern can be observed for all the fluid depths considered. More specifically, N has maxima along the baroclinic wave jet. The minima, instead, change position according to the fluid depth. Close to the surface, low values of N/f are visible towards the outer wall, whilst close to the bottom they appear at the core of the baroclinic wave. We can also notice that $N < f$ in the regions where there is a minimum. This can be relevant for the propagation of atmosphere-like IGWs. Consequently, if gravity waves are spontaneously generated by the baroclinic wave, then they are therefore able to propagate only in regions where $N/f > 1$. In this case, the spatial distribution of N/f shows how the maxima along the baroclinic jet could become a region where IGWs will become trapped. This might be even more accentuated in open surface laboratory experiments, where we have already seen that the oscillations of N/f are larger and therefore, when compared to the simulations, the region for which IGWs propagation is permitted could be even more pronounced along the baroclinic wave jet. We already observed an example of IGWs patterns suggesting wave trapping along the jet in the barostrat experiment (see section

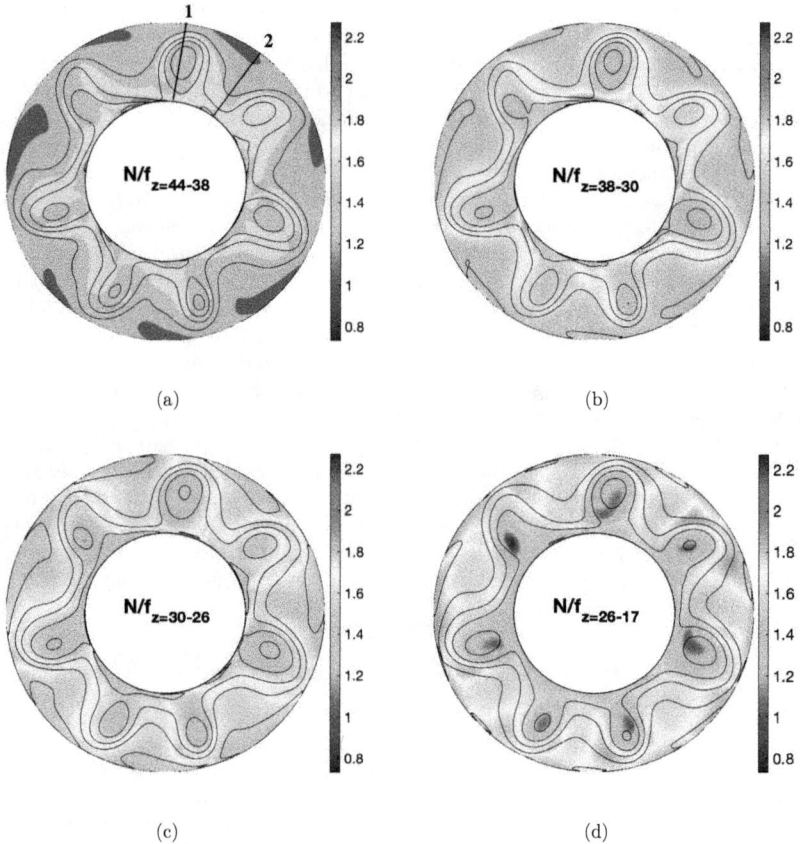

Figure 7.4: Horizontal maps of N/f at different fluid depths (see figure 7.2) from temperature numerical simulations data taken from $z =$1.7, 2.6, 3.0, 3.8 and 4.4 cm. Temperature contour lines are shown. The lines in (a) labelled with 1 and 2 show the positions along which the vertical profiles of figure 7.5 are taken.

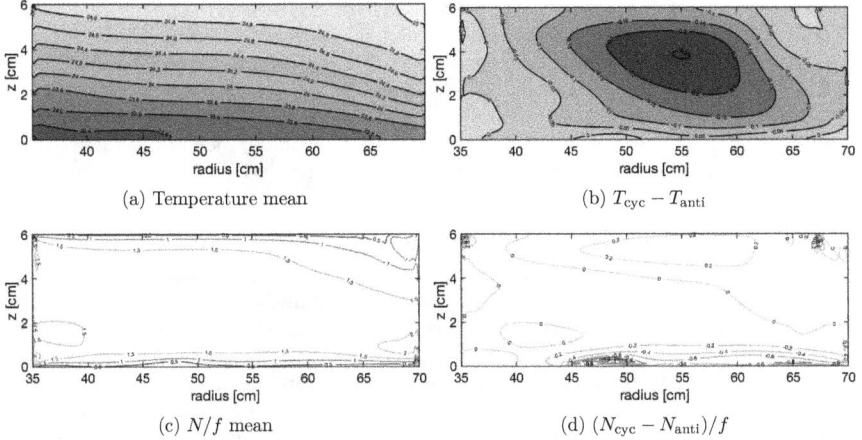

(a) Temperature mean

(b) $T_{\mathrm{cyc}} - T_{\mathrm{anti}}$

(c) N/f mean

(d) $(N_{\mathrm{cyc}} - N_{\mathrm{anti}})/f$

Figure 7.5: Vertical-radial plane of numerical simulations data taken along lines 1 (baroclinic cyclone) and 2 (baroclinic anticyclone) in figure 7.4(a): temperature mean with contour interval 0.2 °C (a), temperature difference with contour interval 0.05 °C (b), N/f mean with contour interval 0.5 (c), and N/f difference with contour interval 0.2 (d).

6.3.1 figure 6.9). In that case, the small-scale waves have been observed to propagate mainly due to advection along the baroclinic wavefront, showing characteristics typical of wave capturing (Bühler and McIntyre 2005). Therefore, we can expect to observe similar patterns also for the atmospheric-like experiment. In summary, we found that the baroclinic jet shapes the background buoyancy frequency N. The distribution of N has a large effect on gravity waves ($N > f$) but not on inertial waves ($N < f$). For $N > f$, gravity usually dominates and, even if N/f is one order of magnitude smaller than in the atmosphere, we might capture essential features of atmospheric jet generated waves. For the classical set-up with $N < f$ this is not possible.

Finally, we consider vertical-radial planes of the temperature and N/f taken along two radial lines one in the centre of the baroclinic cyclone and the other in the baroclinic anticyclone (black lines 1 and 2 in figure 7.4(a)). The mean and the difference are computed and plotted in figure 7.5 (a) and (c) and figure 7.5 (b) and (d) respectively. The temperature shows a variability of maximum 1% in the region where cyclones and anticyclone cores are alternating, assuming lower values in the cyclones. During the cyclone, the isotherms show a positive slope at the inner wall and negative slope at the outer wall, indicating a geostrophic jet at the outer and inner wall. N/f shows a stronger variability close to the upper surface and bottom, with a larger value of N in the upper region and smaller N in the lower region in the cyclone and vice-versa in the anticyclone. In the central region of the gap, the variability is zero. This gives a wider IGW frequency band $f < \omega < N$ located in the upper half during the cyclone and near the bottom during the anticyclone. It is important to note that the horizontal maps show in general a much stronger total variability for N/f than the one measured in the meridional plane.

Borchert et al. (2014) compared similar plots for the classical configuration (figure 3 small-tank in their paper) and for a larger and shallower water configuration (figure 5 big-tank in the paper). The maximum value of N/f they found for the small tank is equal to 0.5, showing that atmosphere-like IGWs cannot propagate in classical configurations where $\Pi > 1$. When $N < f$ the generation and propagation of more inertia dominated gravity waves is still possible. Indeed, the values of N and f define the range of the intrinsic frequency for IGWs, which satisfy the dispersion relation

$$\omega_i = N^2 \cos^2(\alpha) + f^2 \sin^2(\alpha), \qquad (7.3)$$

where α is the angle of phase propagation relative to the horizontal plane. It follows that for $N < f$ low-frequency waves propagate nearly horizontally, whilst high-frequency waves propagate almost vertically. The opposite is true in the atmosphere, where $N > f$. For this reason, IGWs observed in the classical configuration are expected to have a behaviour more similar to inertial waves rather than gravity waves propagating in the atmosphere.

The profiles by Borchert et al. (2014) for the atmosphere-like tank look qualitatively similar to the ones plotted here in figure 7.5. Although the imposed lateral difference of temperature in Borchert et al. (2014) is much higher ($\Delta_r T = 30\,^{\circ}\text{C}$) than what we used (because of the technical reasons mentioned in section 3.3.3), the resulting N/f reaches a maximum value of 4.5 at mid-depth, about two times greater than the maximum value observed here. Except for the top and bottom few millimetres, N/f is always greater than one in the vertical domain. Therefore, atmosphere-like IGW propagation is possible in the entire fluid bulk also for the experimental configuration.

In this section, we investigated in detail the vertical and radial temperature fields using data recorded in the laboratory experiment and the numerical simulations. In general, even though some discrepancies can be expected because of the approximation made in the numerical simulations of an adiabatic and wind stress-free surface, a good quantitative and qualitative match has been observed between the two. What we observed in the experiment, is that the two temperature gradients along the vertical and horizontal direction do not match. Hide (1967) introduced the factors σ_z and σ_r as a measure of the vertical and horizontal temperature gradient respectively and found that $\sigma_z > \sigma_r$ for the classical configuration. Our studies show instead that this ratio reverses, i.e. $\sigma_z < \sigma_z$ for the atmosphere-like configuration. Furthermore, the two parameters are not constant over the regime diagram and, therefore, the approximation we made in our initial discussion, $Fr \approx Ro_T^{(1/2)}$, must be considered as a first guess only. Since $Bu = \sigma_z Ro_T$, $Fr = (Ro_T/\sigma_z)^{(1/2)}$ follows and because σ_z increases towards the geostrophic turbulence area of the regime diagram, it suggests that for this regime the background conditions come closer to the one considered for Lighthill radiation, since Fr becomes smaller as required. Note that still the wave source, considered to be small compared to the long gravity wave in shallow water, is different for the stratified case. Therefore, caution should be exercised when introducing spontaneous emission into stratified fluids in the context of Lighthill radiation.

The detailed study of horizontal and vertical maps of the buoyancy frequency has shown how atmosphere-like IGWs can, in principle, be generated and propagate over almost the entire fluid domain. However, because of larger variations observed in the laboratory experiment compared to the numerical simulations, we speculate that there are larger regions that are forbidden for the gravity waves to propagate. This could result in waves thermally trapped along the baroclinic jet, where the buoyancy frequency is observed to reach its maximum value. This is confirmed by the observation of waves trapped along the baroclinic jet, where the buoyancy frequency is observed to reach its maximum value. A similar effect of gravity waves captured when propagating through regions of the flow with strong shear has been presented by Plougonven and Snyder (2005), whose numerical simulations highlighted how wave capture might be useful to predict the location and several characteristics of spontaneously generated gravity waves. The observation in the atmosphere of waves with similar characteristics (e.g., Suzuki et al. (2013)) and the fact that captured waves have shorter scales and, therefore, can break so that they are relevant for turbulence and mixing, calls for more investigations. Such observations show again that laboratory experiments could support the numerical simulations offering a realistic background flow without scale-dependent approximations.

7.1.4 First results with the modified configuration

An issue pointed out in the previous section is the discrepancy between the numerical simulations and the laboratory experiments in the temperature, and consequently the N field close to the surface. The study of the effect of the upper boundary condition on the gravity wave generation and propagation is still ongoing research, and not included in this thesis. Numerical simulations with heat flux at the water surface, so that a better resemblance with the condition of the laboratory experiment is obtained, are planned for the near future. An experimental campaign has been run with the modified set-ups shown in details in section 3.3.4, i.e. with an internal metallic wall that allows to reach higher lateral temperature differences and, successively, with a plexiglass lid on top of the experiment to investigate the effects at the upper boundary on the flow. In this section, we report on some preliminary results obtained via temperature measurements taken locally with the ALMEMO temperature sensors at the lateral walls and mid-gap.

With the metallic inner wall, the cooling on the inner side of the annulus is much more effective than the previously used configuration with the plexiglass wall. Thanks to this, a maximum temperature difference $\Delta T \approx 9\,°C$, which is double the value reached with the plexiglass wall, can be reached and the lateral temperature difference remains constant over the entire duration of the experiment. However, because the cooling is stronger than the heating on the external wall, the global mean temperature values are $T \approx 15\,°C$, i.e. $\approx 5\,°C$ lower than the ones measured in the experimental set-up with the plexiglass wall. Despite this difference, we expect the flow to show similar dynamics to the one observed in the original configuration.

Following the study and the conclusions drawn in the previous sections, we concentrated the measurements in regions of the diagram regime where we expect the strongest IGWs activity, i.e. at high Ta and low β numbers. A regime diagram comprehensive of the three laboratory set-ups at the BTU–the small tank, the big tank with the plexiglass walls, and the modified big tank with the inner metallic wall—are shown in figure 7.6 together with the corresponding instability curves calculated after O'Neil. The first two straight lines from the left (with red and blue markers) are the same ones plotted in figure 7.1 and thereafter described, the light blue markers indicate the latest measurement done with the metallic inner wall in the big tank. Although we did not investigate the transition between axisymmetric and wave regime for the new configuration, a good agreement with the plexiglass wall configuration can be

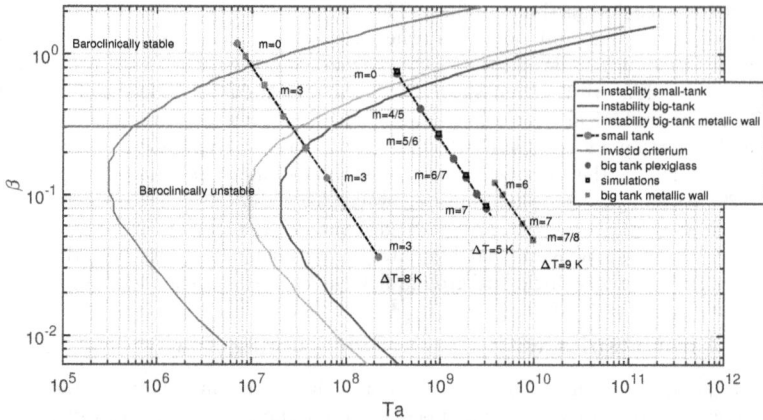

Figure 7.6: Similar plot of the regime diagram in figure 7.1, with added the data from the experiments done with the new configuration with the inner metallic wall. We only investigated areas of the regime diagram with higher rotation rates, focusing where the highest internal gravity wave activity is expected.

observed for higher wavenumbers. Moreover, a wave $m = 8$ has been observed for the highest rotation rate investigated, which is the maximum wavenumber that can set-in for this geometrical construction with a 5 cm narrower gap compared to the plexiglass wall.

One of the main differences observed with the numerical simulations is the more significant amplitude variations of N/f observed close to the water surface for the experiment (shown in figure 7.3 and discussed in the text), which even drop to values lower than 1. To investigate these discrepancies, we repeated the same analysis for the temperature data recorded in the experiments with the two new set-ups. The comparison is shown in figure 7.7, where the plot of N/f for the configuration with the plexiglass inner wall with rotation $\Omega = 0.6$ rpm (a), with the metallic inner wall with rotation $\Omega = 1.1$ rpm (b), and with the metallic wall plus the plexiglass lid fixed at 1 cm above the water surface and with rotation $\Omega = 1.1$ rpm (c) are shown. The red line corresponds to the normalised temperature measured at the upper most point used for calculating N. Figure 7.7(a) is obtained from the same data as 7.3(a) at z_{44-38}, the difference is that the curves have not been interpolated and now show that the minima, occurring in correspondence of the cold core of the baroclinic wave, are regions where $N^2 < 0$. For brevity, the plots for the other fluid heights are not shown here, but a similar behaviour is observed for z_{38-30}, whilst for the two levels close to the bottom $N^2 > 0$ always. This peculiar behaviour observed in the laboratory experiment is of interest since the regions with sign inversion of N^2 might be sources of convective instability, with consequences for the generation of IGWs, as we will explore later in section 7.2.4.

The introduction of the inner metallic wall completely suppresses the drops of N. Moreover, the double peaks corresponding to the entrance and the exit regions of the baroclinic wave are much better visible in the plots, looking more similar to the ones

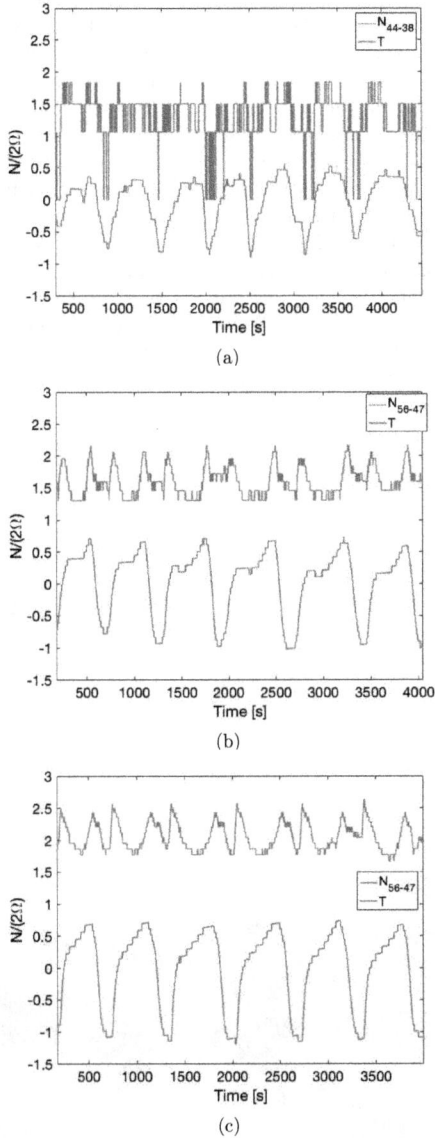

Figure 7.7: Comparison of the time series of N calculated closed to the surface for three different set-ups with the big tank. (a) is for the configuration with the inner plexiglass wall, (b) for the inner metallic wall, and (c) with the metallic inner wall with the additional plexiglass lid above the surface. The red line in all plots show the normalised temperature.

obtained for the numerical simulations (figure 7.3(b) and (d)). Although the rotation rate is almost double the one of the experiments with the plexiglass wall, the mean value of N/f is comparable. Further notice that, since we are interested in studying possible variations at the surface, we positioned the temperature sensors closer to the water surface. For this reason, the data in figure 7.7(b) and (c) have been recorded with sensors positioned $\approx 1\,\text{cm}$ closer to the surface than figure 7.7(a).

The further addition of the plexiglass lid above the surface does not substantially change the behaviour of N/f which looks qualitatively similar to the one without the lid. The mean value shows an increment of ≈ 0.5, possibly indicating that the heat loss at the surface is decreased when the lid is added.

This preliminary analysis of temperature and buoyancy frequency shows that the modifications introduced in the laboratory experiment do seem to converge towards a closer similarity with what was observed in the numerical simulations in terms of the conditions for generation and propagation of gravity waves. However, due to time limitation this has not been worked out further in the thesis but will certainly be part of future work.

7.2   Small-scale waves

We now move towards the investigation of the small-scale waves observed in the rotating annulus. In this section, we first analyse the observed waves, where they are located with respect to the baroclinic jet and their properties, such as propagation velocity, wavelengths and frequencies. After this comprehensive description, we search for propagation and generation mechanisms that can explain the wave properties. Finally, similarities with the observations of gravity waves in the atmosphere are considered.

7.2.1   Properties of the waves

Internal gravity waves, due to their small-scale and short temporal scale, are difficult to observe directly in the total velocity field. Moreover, IGWs motion is coupled to the large scale balanced flow, which makes it necessary to separate both in order to observe the signature of gravity waves. To achieve this separation, we calculated the horizontal divergence from the velocities measured via PIV techniques. This quantity is indeed frequently used in numerical simulations (e.g. O'Sullivan and Dunkerton (1995) and others) and atmospheric observations (e.g. Wu and Zhang (2004), Dörnbrack et al. (2011), and Khaykin et al. (2015)) as a dynamical indicator of IGWs. The limitation of this method is that the horizontal divergence field that is assumed to represent the unbalanced part of the flow might still contain parts of the balanced motions. More sophisticated methods involving a modal decomposition of the full field have been applied by Hien et al. (2018) for a linear system and by Kafiabad and Bartello (2017) and Chouksey et al. (2018) for nonlinear systems to separate balanced and unbalanced fields. Unfortunately, these diagnostic tools require the full three-dimensional velocity field and are, therefore, often impossible to use for laboratory data or observations that generally do not give the full three dimensional information.

A snapshot of the horizontal divergence calculated from the horizontal velocity components, measured at a fluid height $z = 5\,\mathrm{cm}$, is plotted in figure 7.8 in the inset on the left. The plot is superimposed to the surface temperature map reconstructed from the IR camera measurements so that it is easy to observe the position of the small-scale features with respect to the large-scale baroclinic wave. These are concentrated along the jet front and exhibit wavelike structures with the wave crests perpendicular to the flow. They appear irregularly, propagate within the jet stream, are mostly advected by it and then dissipate; we refer to this irregular behaviour as intermittency. A rough estimation of the zonal and meridional horizontal wavenumbers can be done graphically from figure 7.9(a) and are $\lambda_x = 5\,\mathrm{cm}$ and $\lambda_y = 4.4\,\mathrm{cm}$, in the cartesian system of reference. These characteristics, observed for the particular experiment presented here, are very consistent and commonly observed in most of the experiments done with the set-up considered, regardless of the rotation rate of the tank.

The observation of small-scale waves along the jet shows a similarity with the many observations which have identified jets and fronts as significant sources of IGWs in the atmosphere. In particular, observations but also numerical models over the years have consistently shown that the jet exit region and also, although less frequently, at the entrance region are favoured locations for large-amplitude IGWs (see, e.g. Plougonven and Zhang (2014b) and references therein, Dörnbrack et al. (2018), and von Storch et al. (2019)).

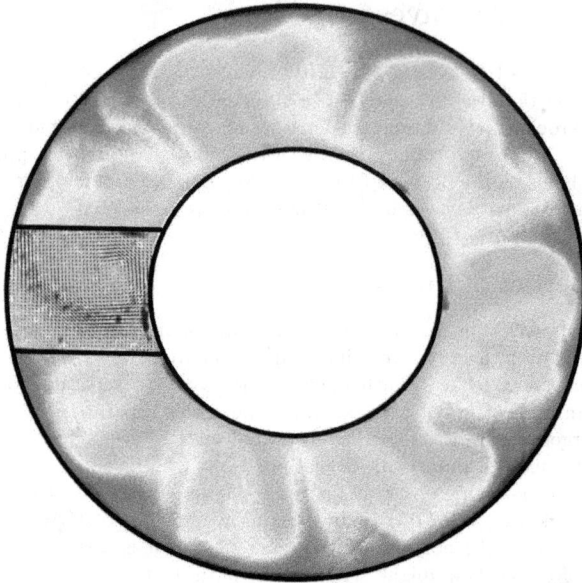

Figure 7.8: Combined plot of the surface temperature reconstructed from measurements done with the infrared camera, and the velocity field (black arrows). The horizontal divergence (colours) measured by PIV in the small portion of the annular gap on the left. Small-scale wavelike features are recognisable in the horizontal divergence field; they appear along the jet and propagate with it.

An enlarged plot of the horizontal divergence, the same used for the composite plot in figure 7.8, is displayed in figure 7.9(a). Here, the velocity vectors are more visible, the axis in physical units are added, and the colorbar shows the divergence amplitude in s^{-1}. Furthermore, the signal-to-noise ratio is given in figure 7.9(b). It has been obtained as described in section 6.3.1 for the barostrat experiment. This plot shows a distinguishable signal along the jet, associated with the wavelike features; furthermore, high values can also be observed close to the inner and outer walls. Since we noticed during the data acquisition that at the walls the light of the laser beam is reflected creating a very bright area nearby, the inner plexiglass wall has been covered with a black coating. This dark covering suppresses the reflection of the laser beam, and therefore a further set of experiments with this modification has been done to investigate whether the signal at the walls is related to physical features or the reflections. The results have shown a significant decrease in the signal along the wall, which indicates that the signal is most likely due to the light reflection that prevents a clear image of the particles in those areas. Therefore, the areas at the boundary walls need to be treated carefully during the data analysis, since the physical features there are most likely hidden by the significant errors in the PIV measurements.

As already done for the barostrat experiment (see section 6.1), we can consider the error on the PIV given by UVmat (see section 4.2.2.1). For the data in figure 7.9 we have $rms = 0.3 - 0.35$ pixels for a typical displacement of 5 pixels. This gives an estimation for the PIV error of $6\% - 7\%$. The percentile of excluded vectors is less than 0.5%.

After having visually identified the wavelike features in several snapshots of the horizontal divergence, we now want to correlate their occurrence with the temporal evolution of the baroclinic wave. To do so, we consider an arc at mid-radius (red line in figure 7.10(a)) and linearly interpolate the radial velocity component and the horizontal divergence along this line as a function of time. The linear interpolation is necessary since the data are distributed on a regular cartesian grid and therefore the points along the arc are not always intersecting with this grid.

In figure 7.10 the plot (b) shows the evolution in time of the mean radial velocity component. One can easily notice that three fronts are travelling through the window over the total measurement time. This happens because the baroclinic wave propagates prograde, and we measure in the co-rotating system. Negative slope corresponds to the exit region of the jet and positive slope to the entrance of the jet.

Plot (b) at the bottom of figure 7.10 shows the horizontal divergence in a time-space plot, also called a Hovmöller plot. The small-scale features are distributed along oblique lines, whose slope has been indicated on the right of the plot by the dashed line. These regions correspond to the baroclinic wavefront, and their slope is the drift speed of the wave itself. Therefore, it can be seen that the small waves are continuously emitted along the jet and propagate with it. This behaviour was already observed for the barostrat experiment in figure 6.12. Similarly, the small-waves do not seem to propagate away from the jet. We will later investigate the possible generation and propagation mechanisms that can help in explaining these particular features.

A similar analysis has been done by Lovegrove et al. (2000) for small-scale waves observed in the shear-driven version of the baroclinic annulus. Their Hovmöller plot shows many features in common with ours. The small-scales are embedded in the front and propagate faster than the drift speed of the baroclinic wave. This can be deducted by the steeper inclination of the short lines (one is indicated with the solid black line to help visualisation) in figure 7.10(b) compared to the flatter diagonal ones crossing the whole Hovmöller plot (dashed black line) and indicates the typical velocity along the front.

7.2.2   IGWs dispersion relation

To further investigate the properties of the observed wave train structures, we perform an analysis in Fourier space. The aim is to identify the frequencies of the waves and see if they fulfil the IGWs dispersion relation (1.1) by spanning the range $f < \omega < N$. A two-dimensional fast Fourier transform (see section 5.2.3 for a detailed description) is applied to the horizontal divergence time series to obtain plots in the frequency-horizontal wavenumber space. To find the wave properties it is important to do a "following the wave" approach in the sense that we choose a line which always lies along the wavefront. The main challenge here is that, since the baroclinic wave travels prograde to the rotation of the tank and PIV data are recorded in the co-rotating system, the portion of the jet visible in the domain will always change over time. Therefore, to be able to follow the

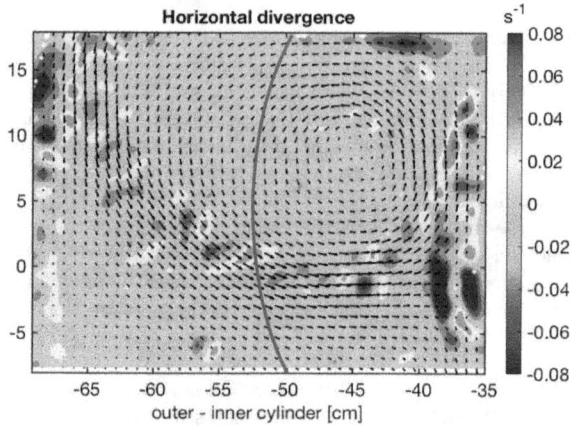

(a) Horizontal divergence

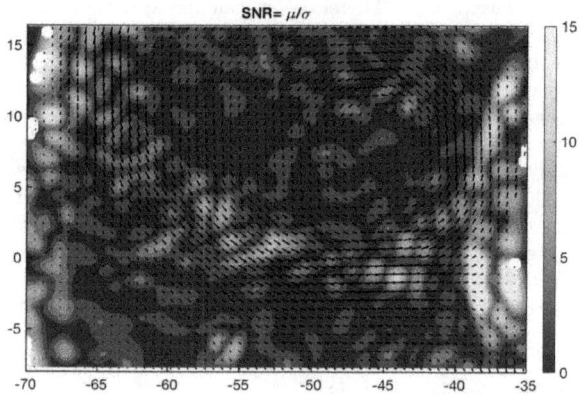

(b) Signal to Noise Ratio

Figure 7.9: (a) Plot of the horizontal velocity field (arrows) and the horizontal divergence (colours) measured with PIV at $z = 5\,cm$. The red curve shows the position along which the data shown in figure 7.10 are taken. (b) plot of the signal to noise ratio $SNR = \sigma/\mu$.

(a)

(b) Mean radial velocity component

(c) Hovmöller plot of the horizontal divergence

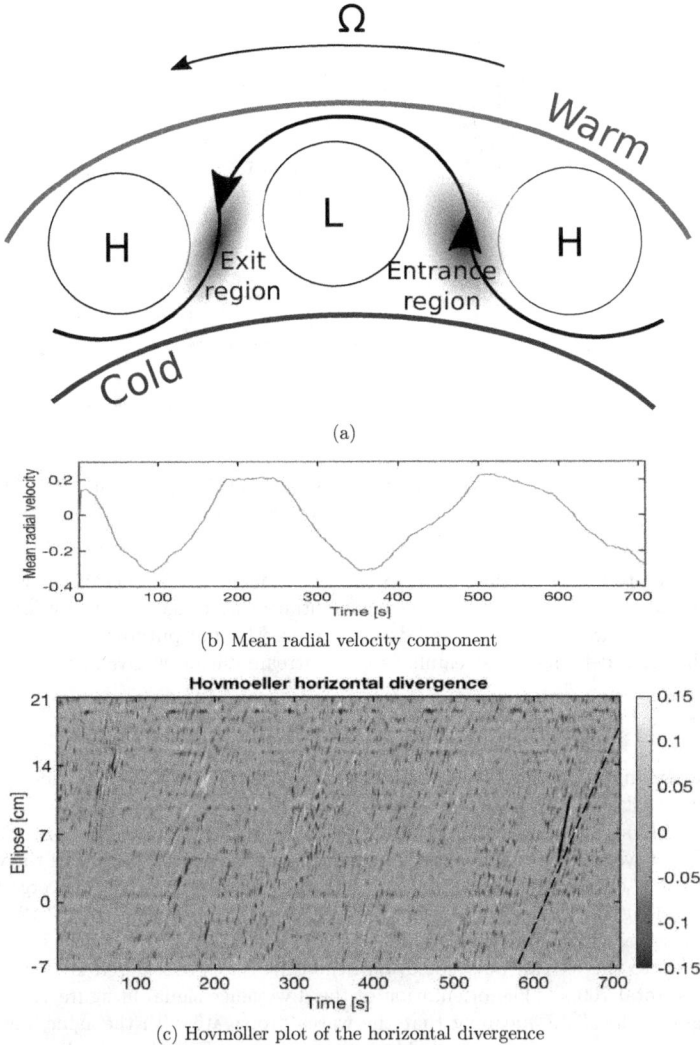

Figure 7.10: (a) sketch of the position of the entrance and exit region (gray shadows) along the baroclinic jet in the rotating annulus. A sector of the experiment is represented and the red and blue lines indicate the warm external and cold internal walls respectively. The circles labelled with L and H show the low and high pressure regions around which the jet meanders form. (b) evolution in time plot of the mean radial velocity component. The positive slope of the radial velocity correspond to the entrance region of the jet, whilst the negative slope is the exit of the jet. (c) horizontal divergence space-time (Hovmöller) plot. A stronger signal in plot (c) is visible in correspondence to these two regions of the baroclinic jet. The dashed line indicates the propagation of the inertia-gravity waves, whilst the solid black line shows the typical velocity along the front.

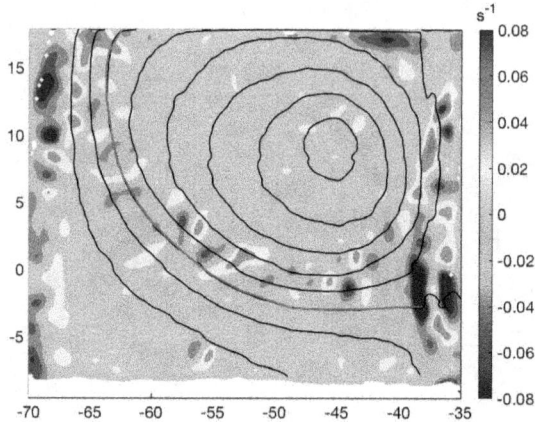

Figure 7.11: Horizontal divergence and streamlines BTU experiment. The streamline marked in red is the one along which plot (a) in figure 7.12 is obtained.

front, we need to find a physical quantity which can identify the front of the wave. Since the streamlines are, by definition, curves instantaneously tangent to the velocity vector of the flow, they are the perfect candidate to be used for our purpose.

The function defining the streamlines is the streamfuncion Ψ given as

$$u = \frac{\partial \Psi}{\partial y} \quad \text{and} \quad v = -\frac{\partial \Psi}{\partial x}. \tag{7.4}$$

Once the stream function is known, the streamlines are straightforwardly identified by a unique constant.

Figure 7.11 shows the streamlines and their paths with respect to the horizontal divergence. The curve marked in red is the streamline along which the maxima of the velocities occur and it is chosen as the representative of the jet. The horizontal divergence is interpolated along this curve, and the Hovmöller plot is obtained (see figure 7.12(a)). Note that for this analysis the streamline is kept fixed so that the evolution of the gravity waves with respect to the jet can be studied, and the time series considered is short (50 s are used of the total 700 s). The orientation of the divergence signal in figure 7.12(a) shows no changes in the slope, meaning that the waves propagate with the same angle to the jet and with a constant phase speed. Moreover, the signal fades in time showing that the strongest waves are embedded in the jet since when the jet is far from the streamline the signal decreases.

From the data used in figure 7.12(a), we can finally proceed to the Fourier space (see section 5.2.3 for more details about the two-dimensional Fourier transform) and analyse the horizontal wavelengths and frequencies plotted in figure 7.12(b). On the x-axis, there is the horizontal wavenumber, which is the resultant of two components: one along the jet and one perpendicular. On the y-axis, the observed frequencies ω_M are plotted. The first feature that can be noticed is an asymmetry in the distribution of the peaks, showing more a prominent signal on the right-hand side of the plot, i.e. for positive wavenumbers.

The explanation for this is that the measured frequency is Doppler shifted, which means that it is the resultant of the intrinsic frequencies of the gravity waves summed to the one at which the flow propagates

$$\omega_M = \omega_i + \vec{U}_0 \cdot \vec{K}_H. \tag{7.5}$$

According to (7.5), the asymmetry can be interpreted as a preferential direction of propagation, which in this case is in the same direction of the mean flow. For waves propagating against the flow, one would observe the negative branch. An example of gravity waves propagating against the flow was found in the barostrat experiment (see figure 6.14).

The white and red lines are the Doppler shifted dispersion curves for IGWs calculated with

$$\omega_i^2 = \frac{N^2 K_H^2 + f^2 n^2}{K_H^2 + n^2}, \tag{7.6}$$

where $N = 0.2\,\mathrm{rad\,s^{-1}}$ is the Brunt-Väisälä frequency (measured with the temperature sensors placed along a vertical line), f is the Coriolis frequency, K_H is the horizontal wavenumber, and n is the vertical wavenumber. The vertical wavelength is not known since we only measure in horizontal planes. The possible variations of λ_z have been tested in the range $0.1\,\mathrm{cm}$ to $1\,\mathrm{cm}$, and all choices are matching the peaks in the plot. The horizontal wavenumber is assumed to vary in the interval $-10\,\mathrm{cm^{-1}}$ to $10\,\mathrm{cm^{-1}}$, which is the range spanned by the measured waves. The frequencies are corrected, in order to match with the observed ones, by the Doppler shift calculated considering $U_{min} = 0.2\,\mathrm{cm\,s^{-1}}$ and $U_{max} = 0.4\,\mathrm{cm\,s^{-1}}$, which are the maximum and minimum velocity magnitudes measured along the jet. The asymmetry between the positive and negative wavenumber regions is to be attributed to the Doppler shift. The region with the most significant amount of energy is included within the two dispersion curves, indicating that it is related to gravity waves.

The same analysis can be extended to the entire time series, this time considering a variating streamline that follows the jet over time. The details about this procedure and the results are presented in the appendix A.4. The excellent agreement obtained with the two methods shows that most of the globally measured energy belongs to the gravity waves propagating along the jet.

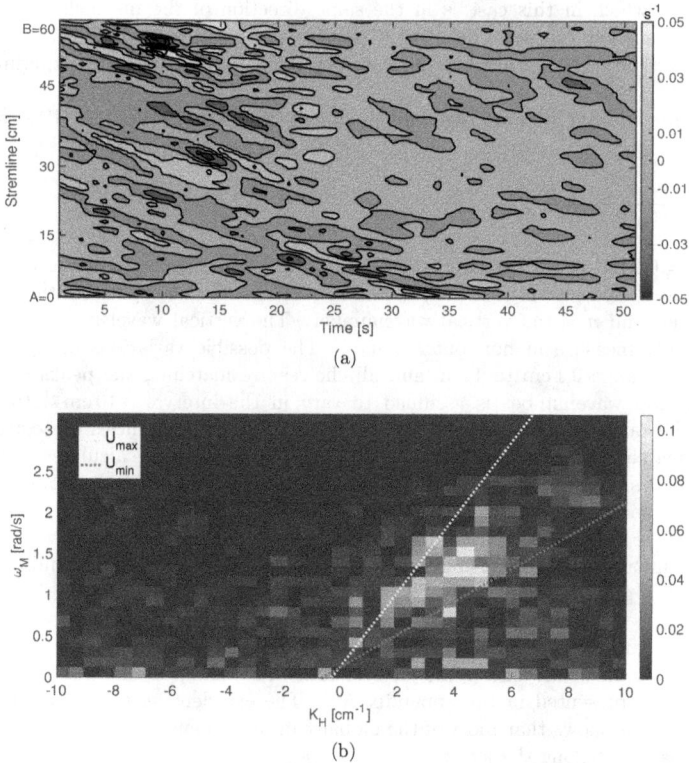

Figure 7.12: Analysis of the frequencies and wavenumbers along a fixed streamline. (a) Hovmöller plot of the horizontal divergence for $1s \leq t \leq 50s$, and (b) 2D fft of (a) with IGWs dispersion curves calculated by using (7.6). The curve in white is doppler shifted using the maximum velocity $U_{\max}$ measured along the jet and the red one with the minimum velocity $U_{\min}$.

7.2.3 Wave amplitude scaling

After having investigated which regions of the flow show the most prominent signature of waves, it is of great interest to seek a possible dependency of the gravity wave amplitude on the bulk flow properties. The investigation of a proper scaling, i.e. of the connection between the waves and the large-scale flow, is the core of any parameterisation, which has vital importance for weather forecast and climate modelling. One big problem in the context of parametrisation of non-orographic and non-convective gravity waves is the lack of a clear understanding of what is the balanced part of the flow and what is the unbalanced part which is still missing and debated (Dritschel and Viúdez 2003, Warn et al. 1995). With the laboratory experiments, we have access to a limited number of variables and also only to some regions of the flow. For this reason, with the experimental data alone the separation of balanced and unbalanced flow is out of reach. However, in combination with theoretical research, the issue of imbalance and a corresponding sound scaling analysis can be addressed, and findings can be verified in the laboratory.

The scaling of gravity waves with the Rossby number Ro and the Froude number Fr is a current topic which is addressed by both theoretical and experimental research. From the many studies of spontaneous emission of IGWs in the regime $Ro << 1$, there seems to be consistency in a decrease of the wave amplitude with decreasing Rossby number (see Williams et al. (2008) and references therein). There are two theories, and consequently, two scalings proposed in the literature, the first considers the ageostrophic effects that appear at the first order in Ro for an asymptotic expansion of the governing equations (Pedlosky 1982), and the second is a low-order model analogous to the coupled spring-pendulum system (see section 2.6.1.3). These two show different amplitude dependencies on Ro, the first scaling as Ro^{β} with $\beta \geq 1$, while the second goes as $Ro^{\gamma} \exp(-\alpha/Ro)$ with $\gamma = -2$ and $\alpha \geq \pi/2$ after Vanneste and Yavneh (2004) or $\gamma = -1/2$ after Plougonven and Snyder (2005). The power-law dependence has been obtained in numerical simulations of vortex dipoles by Snyder et al. (2007) with typical values of $\beta \approx 4$ and by Wang et al. (2009) who observed $\beta \approx 6$. Such dependency has also been found for small-scale waves observed in laboratory experiments by Williams et al. (2008) with $\beta \approx 1$, although the generation mechanism has not been clarified and the regime reached by the experiment is not consistent with the Lighthill theory of wave radiation. From what we just discussed, showing the expected weakness of gravity wave emission for flows in a balanced state with $Ro << 1$, one might wonder whether these waves are significant and can be measured in the real atmosphere. To this regard, Vanneste and Yavneh (2007) write that spontaneously emitted waves can become important suddenly as the Rossby number increases towards 1 and beyond. This, indeed, seems to be the case for our results, for which a measurable signal of gravity waves is seen only in regions where $Ro > 1$, as highlighted in figure 7.15.

In the following, we investigate the observed occurrence of gravity-waves as a function of the thermal Rossby number. To this purpose, five experiments have been run with the big-tank where the same lateral temperature difference $\Delta T_H = 4.8\,^{\circ}\mathrm{C}$ was imposed and the rotation rate was varied. The experiments, therefore, have different thermal Rossby number, which we recall is proportional to the temperature difference ($Ro_T \propto \Delta T_H$) and inversely proportional to the rotation rate squared ($Ro_T \propto 1/\Omega^2$).

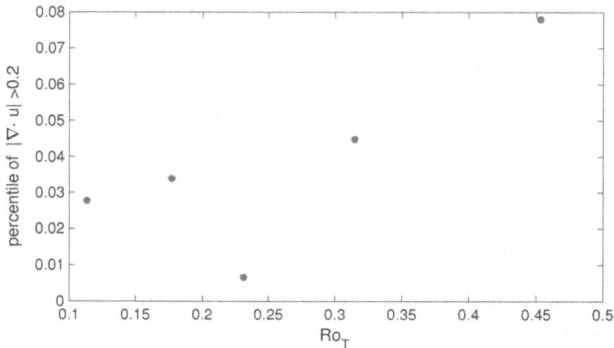

Figure 7.13: Plot of the 'percentile' of IGWs (larger than the threshold $|\nabla \cdot \vec{u}| > 0.2\,\mathrm{s}^{-1}$) as a function of the thermal Rossby number Ro_T. See text for more detail.

The investigation aims to see whether some scaling proposed by the theoretical models can be experimentally verified. We already discussed that the horizontal divergence is a good indicator for gravity waves activity. If we want to investigate the amplitude of gravity waves, however, the horizontal divergence is not the best quantity as it also includes part of the balanced flow. Without a feasible method to separate the balanced from the imbalanced part, the amplitude of gravity waves in the experiment seems to be unavailable.

What we can do is use the horizontal divergence as an indicator of gravity waves and investigate whether they are more or less prominent at different considered thermal Rossby numbers. To get a quantitative measurement and compare the amount of gravity waves occurring for the different experiments, we proceed in the following way: for each experiment the horizontal divergence is calculated over the entire domain at every time sampled. Afterwards, the Hovmöller plot along a middle-radius arc is considered, similar to that shown in figure 7.10(c). From this, a minimum threshold $|\nabla \cdot \vec{u}| > 0.2\,\mathrm{s}^{-1}$ is set and only values of the divergence above such threshold are further considered. Then, the total number of points in the domain having divergence values above the threshold are counted and normalised by the total number of points in the domain. This 'percentile' quantity is what we consider as a measure of the gravity wave activity for each experiment. Figure 7.13(a) shows the 'percentile' of horizontal divergence $|\nabla \cdot \vec{u}| = 0.2\,\mathrm{s}^{-1}$ as a function of the thermal Rossby number Ro_T. Except for one outliner point, the trend is that enhancement of gravity wave activity is observed for increasing Ro_T.

Although we have investigated only a limited number of cases and we do not have enough points to propose a scaling relation for gravity waves with the Rossby number, these first results show the potential that our laboratory experiment has for finding proper parametrization. More experiments are planned for the future to provide more data and find for a sound scaling that could help to validate one of the theoretical models. As we have already discussed, the scaling of gravity waves with Ro and Fr is ongoing research and not available yet for continuously stratified fluids. The knowledge of the proper scaling would give insights on the connection between the waves and the large-scale flow, and consequently would allow for parameterisation of non-orographic and non-convective gravity waves in weather forecast and climate modelling.

7.2.4   Generation mechanism

Several different internal generation mechanisms might be triggering the gravity waves observed in the tank. Although our data are partial, since we do not perform global PIV measurements and we only have the two horizontal velocity components, we can investigate the main candidate mechanisms and see if we can at least exclude some. In this section, we look at four possible generating mechanisms, namely shear instability, lateral boundary instability, convection, and spontaneous imbalance.

Kelvin-Helmholtz shear instability   Similarly to what was done for the barostrat experiment in section 6.3.2, we investigate whether shear instability (see section 2.6.1.4 for an explanation of the instability mechanism and the Richardson number criterium) can occur in the atmosphere-like experiment. To do so, we consider data in the upper part of the water column, close to the surface and calculate the ratio between the buoyancy frequency N and the vertical shear of the flow. Since N is derived by the temperature measurements taken locally along points on a vertical line at the middle gap width, we can only verify whether the shear instability can or cannot occur at this point. Moreover, since the PIV measurements are not performed simultaneously, the vertical shear is calculated for each fluid depth independently, without considering possible phase shifts between the waves at different depths. Keeping in mind these limitations, we proceed with the verification of the shear instability criterium. This is done by taking the mean azimuthal velocity in a point positioned in the middle of the gap at heights $z = 5$ cm and $z = 4$ cm over time. Since the acquired data do not cover an entire wave period, the mean is calculated over two wave crests for both heights to ensure consistency. Then, the maximum and minimum values of N are taken from the data plotted in figure 7.3(a) and the Richardson number (2.60) is calculated. The result is $9.7 < Ri < 39$, a range which is much higher than $Ri_{critical} = 0.25$ and, therefore, rules out shear instabilities for the investigated region of the tank.

The advantage of having directly comparable numerical simulations, which unfortunately are not available for the barostrat experiment, is that they offer insights of the flow over the entire domain and can, therefore, be used to investigate the possible local occurrence of shear instabilities at a given height. A horizontal map of the Richardson number calculated between the same heights considered for the experiment and for a comparable run, is shown in figure 7.14. The range of values is in excellent agreement with the one found for the laboratory experiment, with some maxima close to the inner wall which are outside of the areas that could be investigated experimentally. This map further validates what we already found, i.e. that the shear instability can be ruled out and is not a possible generation mechanism for gravity waves since $Ri > 10$ everywhere in the tank.

Lateral boundary instability   The interaction of the balanced flow with the inner and outer wall of the tank can lead to lateral boundary instability that might be a source for the generation of gravity waves.

A combined simulation-laboratory experimental study on the generation of small-scale waves at the inner and outer walls of the differentially heated rotating annulus has been done by Von Larcher et al. (2018). What they found out is that the instability at the outer

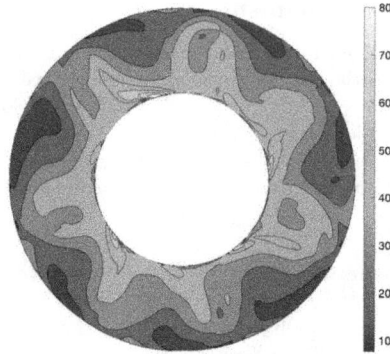

Figure 7.14: Horizontal map of the Richardson number calculated for the numerical simulations between the fluid depths $z = 5\,\mathrm{cm}$ and $z = 4\,\mathrm{cm}$.

warm boundary wall does not interact with the large-scale flow and the small-scale features observed tend to stay trapped at the outer boundary layer. Moreover, those features have frequencies lying outside the ones allowed by the dispersion relation of gravity waves. The situation at the cold inner wall is different. There, the boundary instability does induce gravity waves, which are observed to form in the lower depth region. Evidences of gravity waves generated at the inner boundary have been observed in the differentially heated rotating annulus experimentally by Jacoby et al. (2011), and numerically by Randriamampianina and del Arco (2015). The waves have the characteristic that they start to propagate in a retrograde direction with respect to the large-scale baroclinic drift. Both the facts that the gravity waves observed in our experiments are mostly observed in the upper half depth, closer to the surface, and that those waves move in the same direction of the flow do not agree with what has been reported by Von Larcher et al. (2018).

The thickness of the Ekman viscous layers at the bottom of the atmospheric-like setup is of the same order of magnitude ($\approx 0.5\,\mathrm{cm}$) for both experimental configurations. The thickness of the viscous Stewartson layers at the lateral walls, however, increases of one order of magnitude in the wider annulus, reaching the order of a few centimetres (see table 3.2). A Stewartson layer thicker than the thermal boundary layer would in general point towards a weak feedback between the sidewall temperature forcing and the baroclinic waves since the extent of the waves in the gap is constrained by the Stewartson layer whilst the imposed temperature gradient is handled in the thermal boundary layer (Von Larcher et al. 2018). This would further point to a negligible source of gravity waves due to the lateral boundaries in the atmospheric-like configuration of the rotating annulus.

Convection   As we have seen in section 2.6.1.2, convection is one of the most important non-orographic sources of gravity waves in the atmosphere. Besides the regions within storm clouds and over the cloud tops, where convectively generated gravity waves arise, also at jet/front systems, moist convection is known to possibly force, modify, or enhance gravity waves and have an impact on the evolution and imbalance of the jet itself (Wei and Zhang 2015). Numerical studies by Wei and Zhang (2014) have investigated the effect of introducing convective sources on the emission and propagation of gravity waves from the baroclinic jet and front systems. The convectively generated gravity waves in their system are found to be more prevalent, shorter in scales, and larger in amplitude.

The mechanism is due, more in general in a stably stratified fluid, to a temporary instability of the fluid caused by denser fluid lifted above the lighter fluid. This unstable situation, usually signalled by a sharp localised decrease in the buoyancy frequency N, can cause the overturning of the isopycnals and therefore trigger convective instability that can successively generate gravity waves.

As we have previously seen in figure 7.7(a), sudden decreases of N are observed in the experiment in the middle regions of the baroclinic cyclon. This is an indicator of possible areas where convective instability and gravity wave generation might take place. With the present data, we cannot verify the importance of convection. Future studies with the new experimental set-ups, which we have seen do not reveal areas of convective sources (figure 7.7(b) and (c)), are suggested to solve the uncertainty about the gravity wave generation mechanism.

Spontaneous emission   In section 2.6.1.3 we have exploited the concept of spontaneous emission, which is the mechanism according to which the evolution of well-balanced flows would lead to the emission of IGWs. Indeed, although the flow is balanced and therefore the global Rossby number is small, it can locally grow in localised regions as the baroclinic fronts, and therefore lead to the emission of IGWs. In particular, spontaneous emission is found to be more prominent for $Ro = O(1)$ or larger. In contrast, a $Ro \ll 1$ regime shows a very weak to nearly negligible emission of gravity waves. For this reason, many studies have discussed the use of the Rossby number as an indicator for regions of imbalance in the atmosphere. O'Sullivan and Dunkerton (1995), for example, associated a Lagrangian Rossby number, defined as $Ro^L = (\delta_t V/f)$, with the location of IGWs activity.

Using a similar approach, we can look for a correlation between IGWs and an enhancement of the local Rossby number. In figure 7.15 the Hovmöller plot of the horizontal divergence is displayed one more time with red, blue, and black lines showing the local Rossby number larger than 1, 1.5, and 2 respectively. The local Rossby number is calculated as:

$$Ro = \frac{\zeta}{f}, \qquad\qquad\qquad (7.7)$$

where $\zeta = \partial v/\partial x - \partial u/\partial y$ is the vorticity. The red contour lines show values of the Rossby number larger than 1. It is interesting to notice that they correspond to large values in the horizontal divergence and they are located in the exit region of the jet, which we know is a favourable area for gravity waves emission. This correlation hints to spontaneous imbalance as the generating mechanism.

7.2.5   Propagation and wave capture

After being generated somewhere in the fluid, gravity waves propagate within it. During this process, the characteristics of the background flow can, in particular cases, reorganise

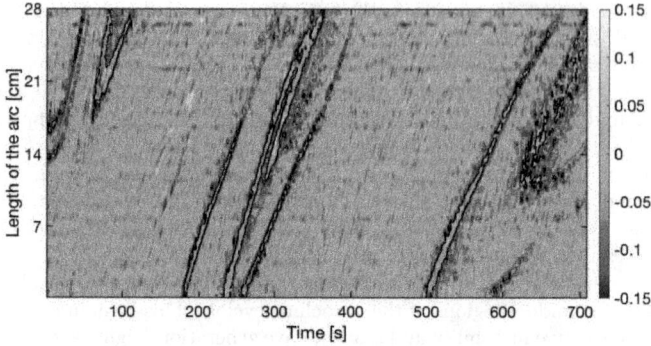

Figure 7.15: Dependence of the horizontal divergence on the local Rossby number maxima. The Hovmöller plot of the horizontal divergence is shown superimposed with red lines contouring the areas where the local Rossby number, Ro (calculated with (7.7)) is larger than 1, blue lines for $Ro > 1.5$, and black lines for $Ro > 2$.

the gravity wave field modifying some properties. In particular, regions with strong horizontal deformation and vertical shear are known for playing a key role in the gravity wave propagation giving rise to a phenomenon called wave capture (Bühler and McIntyre 2005, Plougonven and Snyder 2005). Interestingly, the exit region of the jet is a place where large deformation is observed as a consequence of the deceleration of the flow. Therefore, besides being an interesting region for the generation of waves, the exit region of the jet is also an area relevant for the propagation of waves.

To investigate the wave capture mechanism, we need to look at the properties of the background flow. The method to calculate the stretching and shearing deformations of the flow, that then combine to give the horizontal flow deformation is explain in detail in section 5.2.2. Regions with strong horizontal deformation can be identified by computing the mean deformation defined as

$$\alpha = \frac{1}{2}\sqrt{(u_x - v_y)^2 + (u_y + v_x)^2}, \tag{7.8}$$

and the local contraction axis has the horizontal azimuthal angle θ defined as

$$\tan(2\theta) = \frac{(u_y + v_x)}{(u_x - v_y)}, \tag{7.9}$$

where u and v are the flow components in the horizontal and vertical directions respectively.

The horizontal velocity components, measured with PIV, are smoothed using a Gaussian filter on a square domain of $1\,\mathrm{cm} \times 1\,\mathrm{cm}$ and then the mean deformation and the local contraction axis are computed. Different smoothing window sizes have been tested and the contraction axis does not show significant changes nor in the direction nor in the intensity for windows up to 4 times the one chosen here. The obtained contraction axis are depicted by the oriented lines in figure 7.16 and compared with the horizontal divergence. When wave capture occurs, the horizontal wave vectors tend to align with

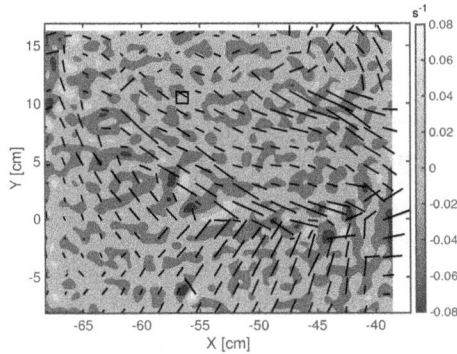

Figure 7.16: Horizontal divergence in colours and the lines are the deformation (amplitude and angle given by (7.8) and (7.9) respectively). The square shows the smoothing window used on the data of dimensions 1 cm × 1 cm.

the contraction axis of the flow, whereas the tilt of wave vectors tends to converge to a value given by the ratio of vertical shear and deformation (Wang et al. 2009). It can be noticed that in the shown plot the wave trains have, indeed, horizontal wave vectors that tend to orient along the contraction axis and occur in regions where the deformation is larger.

Wave capture has been observed in numerical simulations by Plougonven and Snyder (2005) and this process influences the properties of spontaneously emitted gravity waves substantially. According to their findings, larger amplitude waves whose wave crests are perpendicular to the flow and propagating with the local flow speed can be expected if they undergo to wave capture. This propagating mechanism may or may not occur depending on the speed at which the wave packets travel through the section of the flow with a large deformation (Wang et al. 2009). It is clear from this section that if wave capture does occur and modify the characteristics of the waves, it might become more difficult to establish where and how they are generated.

In the previous sections, we investigated the possible generation and propagation mechanisms responsible for the waves observed in the laboratory experiment. With the help of the numerical simulations, we were able to rule out shear instability from the candidate mechanisms, and following an analysis for the thickness of the lateral boundary layers, we suggest that instabilities at the boundaries are not responsible for the gravity waves observed. This leaves us with two generation mechanisms: convection and spontaneous imbalance. The characteristics of the gravity waves emitted by these two processes differ with those generated from convective cells that have short horizontal wavelengths, long vertical wavelengths, and correspondingly high intrinsic frequencies, whereas the spontaneously generated ones have short vertical wavelengths and near-inertial frequencies (Plougonven and Zhang 2014b).

The waves we observe in the experiment have characteristics matching the ones of convectively generated waves or captured waves. In the latter case, gravity waves could be spontaneously generated and then propagate through regions of the background flow with strong deformation and vertical shear. The propagation shapes the gravity waves reducing their horizontal scales, enhancing their amplitudes, and aligning them perpendicularly to the jet.

With the data we analysed both options seem possible, so that it is not possible to discern whether the waves observed are generated convectively or spontaneously and then undergo wave-capturing. Finally, we want to point out that in either case, since both convection and wave capturing predict a wave amplitude larger than the one expected for spontaneously emitted waves, it might be that those are below the resolution of our data, and therefore we simply cannot observe them with the actual measurement instruments used.

7.2.6   Comparison with numerical simulations – small-scale waves

In sections 7.1.2 and 7.1.3 we have presented a detailed investigation of the temperature and of the buoyancy frequency fields. Despite an overall agreement between the laboratory experiments and the numerical simulations, some discrepancies could be observed and we pointed at them as possible reason for differences in the characteristics of the gravity waves observed in the two systems. In this section we show the results obtained for the numerical simulations and discuss them in comparison to the ones observed for the laboratory experiments, whose characteristics have been explored in details in the previous sections.

Two snapshots representative of the typical horizontal divergence fields of the numerical simulations are shown in figure 7.17. The simulations corresponding with the plot shown on the left hand side are the same as the ones used for the investigation in section 7.1.2 and are run with geometry and parameters comparable to the laboratory experiments, where the lateral temperature difference is $\Delta T = 4.8\,°\mathrm{C}$. The simulations corresponding with the plot on the right-hand side are run for a wider gap geometry $b - a = 50\,\mathrm{cm}$, a larger difference of temperature $\Delta T = 30\,°\mathrm{C}$, and consequently a larger rotation rate $\Omega = 0.76$ rpm (for more details, see Hien et al. (2018)).

The two fields appear very different from one another. The most evident difference between the two plots is that for lower temperature difference one main wave packet forms near the temperature front (jet-exit region) and some structures near the outer wall. On the contrary, for a higher ΔT four wave packets develop in the flow. Qualitatively, both look rather different from what is observed in the laboratory, with only WP1 in figure 7.17(b) resembling the dimensions and orientation of the gravity waves. Moreover, the amplitudes in the simulations are half to one order of magnitude smaller than the ones measured in the laboratory.

We speculate on some possible reasons for these discrepancies. The cause for the observed differences could be that the waves are generated by different mechanisms, with the waves in the laboratory being generated by convection and the ones in the simulations by spontaneous emission. This hypothesis is sustained by the comparison of the buoyancy frequency field, which has shown regions where convective instability can occur for the laboratory data, but not for the numerical ones (see section 7.1.3). The experiments with the new set-up, which we have shown in section 7.1.4 do not have convectively unstable regions, could help in future studies to verify whether this hypothesis is correct. Another possibility is that the gravity waves in the laboratory undergo to wave capture during their propagation whilst the gravity waves in the simulations are not subjected to capturing. This second hypothesis would leave open the possibility for the generation mechanism to be the same, but some differences in the background flow might favour the wave capture

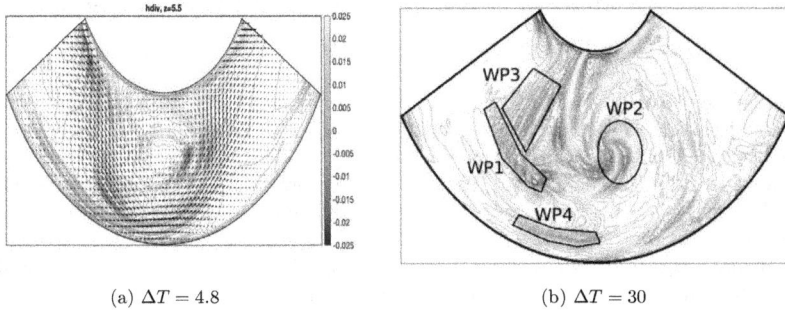

(a) $\Delta T = 4.8$ (b) $\Delta T = 30$

Figure 7.17: IGWs observed in the numerical simulations. (a) horizontal divergence and velocity vectors at $z = 5$ cm for the numerical simulations with total fluid depth $d = 6$ cm $\Delta T = 4.8\,^{\circ}\text{C}$, $\Omega = 0.6$ rpm. (b) unbalanced part of the horizontal divergence at $z = 3$ cm for the numerical simulations with total fluid depth $d = 4$ cm $\Delta T = 30\,^{\circ}\text{C}$, $\Omega = 0.76$ rpm (taken from (Hien et al. 2018)). Contour values ranging from -0.05 s^{-1} to 0.05 s^{-1}.

mechanism in the laboratory experiment and not in the numerical simulations. Finally, the discrepancies between the upper boundary conditions for the experiment and the simulations (discussed in detail in section 7.1.3) could lead to the dissimilarities observed in the gravity wave field.

Future studies are needed to validate the ideas proposed here. For example, experiments with a rigid upper lid (see section 3.3.4.2) and numerical simulation with upper boundary condition allowing heat exchange are undergoing research and will be used in the future to investigate the impact of the upper boundary condition on the gravity wave properties.

7.2.7 Energy spectra and comparison with the atmosphere

After having explained in detail the properties of the observed gravity waves and analysed the possible wave generation and propagation mechanisms in our experiment in the
previous sections, we now want to take a step back and get a more global picture of wave
interactions and energy exchange at different scales. The study and interpretation of
physical phenomena responsible for the energy at the mesoscale range is of fundamental
importance. As Nastrom et al. (1984) wrote: 'Before this problem of parameterisation
can be solved, it is necessary first to observe and to understand the mesoscale portion of
the kinetic energy spectrum.'

By investigating the energy distribution among different phenomena in our experiment,
we firstly want to make a connection between what is observed in the real atmosphere and
our experiment so that the utility of this experiment for atmospheric studies can be highlighted. After such an analogy has been established, we can investigate the contribution
to the total energy from the various phenomena at different scales.

7.2.7.1 Atmospheric spectra

Throughout this thesis, we have discussed the atmospheric motions and how they can be
divided into synoptic-scale balanced motions, including baroclinic waves, and mesoscale
unbalanced motions, among which there are gravity waves. These two types of motion
have very distinct characteristics, and their scales are usually well separated. This is
obvious from atmospheric data and in particular from the energy spectra obtained via
velocity and temperature measurements.

In the 1980s Nastrom and Gage (1985) provided the first comprehensive atmospheric
spectra of zonal and meridional velocities and temperature, measured near the tropopause
level by commercial aircrafts. In their study, the authors noticed that the three spectra
exhibit two distinct power law dependencies upon wavenumber in the form $\mathcal{P} \propto k^{-p_k}$,
where the typical measured values are $p_k = 3$ in the synoptic scale range (for wavelengths
between $500\,\mathrm{km}$ and $3000\,\mathrm{km}$) and $p_k = 5/3$ in the mesoscale range (for wavelengths
smaller than $500\,\mathrm{km}$). These well-known atmospheric spectra, which have been confirmed
with the analysis of a more recent data set by (Callies et al. 2014), are shown in figure
7.18. The data were collected by instruments mounted on commercial aircrafts over a
period of five years. The altitude considered for the analysis are ranging between 9
and $14\,\mathrm{km}$ and the latitudes between 30°N and 60°N. Although some small seasonal and
latitudinal variations are observed, the spectra show that both synoptic and mesoscale
are characterised by global features, which reflect in the two slopes.

The steep slope corresponding to the -3 power law exhibited by the quasi-geostrophic
balanced flow at synoptic scales, is consistent—and well accepted—with Charney's theory
of geostrophic turbulence (e.g. see Nastrom et al. (1984), Callies et al. (2014), Kafiabad
and Bartello (2018)). For this subrange of the spectrum the energy is transferred from
smaller towards larger scales in the so-called inverse energy cascade.

It appears very clear from figure 7.18 that the slope of the spectra for wavelengths
smaller than $500\,\mathrm{km}$ flattens and becomes close to $-5/3$, suggesting that some different
phenomena could be responsible for this part of the spectrum. Many investigations over
the past years have focused on the mesoscale energy subrange trying to explain what
are the phenomena involved (see, for example, Waite and Snyder (2009), Callies et al.
(2014) and references therein). However, the source of energy at these smaller scales

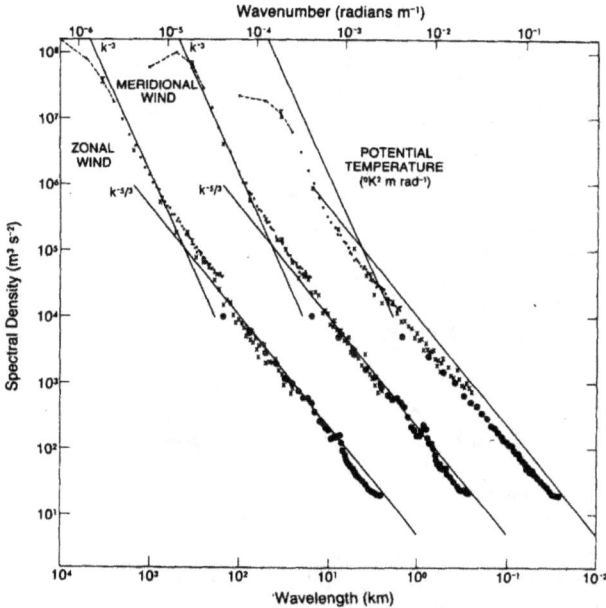

Figure 7.18: Variance power spectra of zonal wind, meridional wind and potential temperature near the tropopause from Global Atmospheric Sampling aircraft data. The spectra for meridional wind and temperature are shifted one and two decades to the right, respectively. Figure taken from Nastrom and Gage (1985).

remains subject of debate. Among the many mechanisms proposed, the most accredited are two turbulent theories: an inverse cascade of small-scale energy, possibly forced by convection (for example studies by Gage (1979) and Lilly (1983)), opposed to a direct cascade of energy from the larger scales (see Cho and Lindborg (2001) and Lindborg (2005)). Together with these two turbulent theories, there is a third possible hypothesis that inertia-gravity waves are, if not completely responsible, at least partially contributing to the energy at the mesoscales (see application on atmospheric data by Callies et al. (2014) of the wave-vortex decomposition method proposed by Bühler et al. (2014)).

7.2.7.2 Experimental spectra

To investigate the energy distribution over the different scale phenomena in the laboratory experiment, we follow the procedure used by (Callies et al. 2014) so that we can compare the result obtained in the laboratory with their findings for data collected in the atmosphere by aircrafts. We calculate the global energy spectra for three configurations of the differentially heated rotating annulus run in our laboratories: the small tank in the classical configuration, i.e. filled with pure water, the barostrat version of the small-tank, and the atmosphere-like tank. After having obtained and compared the energy spectra, we want to investigate whether the energy distribution among different scales resemble the one observed in the atmosphere and if so, seek for possible energy exchanges between the large- and the small-scales. The theory and the wave-vortex decomposition method proposed by Bühler et al. (2014), which is used in the following to separate the contributions from different phenomena to the energy spectra, are explained in section 5.2.3.2.

Since such a method has been developed to analyse one dimensional data measured along flight paths, we shall also select a line along which our data are chosen for the analysis. However, since our PIV field of view is approximatively 1/5 of the total annular gap, by considering the data along a line we do not have a large enough domain to resolve the large-scale waves. A better way to proceed with our dataset, which is obtained by recording videos of long duration, is to investigate frequency spectra. The problem of having frequency instead of wavenumber spectra is not new to the experimentalists since often observations are obtained at single stations. To convert frequency spectra into wavenumber spectra, the Taylor transformation could be used under the hypothesis of frozen turbulence. This conversion is widely used in observations and laboratory experiments (e.g. Benielli and Sommeria (1996)). Furthermore, Gage and Nastrom (1986) compared the very spectra shown in figure 7.18 with frequency spectra obtained from independent measurements and found a good agreement between the two. Their comparison, therefore, validates the use of the Taylor hypothesis.

The horizontal velocity components measured in a cartesian system are interpolated on a polar system so that the components along the radial and azimuthal directions are obtained. Subsequently, as already done many times in this thesis, the data along the mid-radius arc are considered and the power spectral density for each component is calculated $\hat{C}_u$ and $\hat{C}_v$ for every point along the arc. Then average spectra is calculated from all the spectra along the arc so that local effects are smoothed out. The procedure of averaging data over many measurements is also done by Nastrom et al. (1984) and Callies et al. (2014).

The kinetic energy spectra are calculated as

$$\hat{E}_{\text{kin}} = \frac{\hat{C}_u + \hat{C}_v}{2}. \tag{7.10}$$

The comparison of the kinetic energy spectra for the three experiments is shown in figure 7.19, where (a) is the spectra for the classical small-tank, (b) for the barostrat experiment, and (c) for the atmosphere-like experiment. All three experiments are analysed in a respective externally imposed parameters set that led to a regular baroclinic wave regime. In each plot, two straight lines with slope -3 (in blue) and $-5/3$ (in dashed red) are drawn for better visualisation and comparison with the slopes in the atmosphere. Furthermore, in figure 7.19(c), the dashed red line indicated the slope -2, about which we shall discuss later. The classical configuration (figure 7.19(a)) does not resemble the atmospheric features, neither in the large nor in the small scales. The kinetic spectrum is flatter than the two indicated slopes, and shows no clear subdivision between scales. The spectrum obtained with the data from the barostrat experiment (figure 7.19(b)) looks very different and shows a steep slope with no peaks for the largest scales, a peak at the frequency of the baroclinic wave and then a flatter tail for smaller scales. The two slopes are close to the atmospheric ones, although the match is not perfect. One thing that can be noticed is that the baroclinic and larger scales are well resolved, whilst the smaller scale frequencies

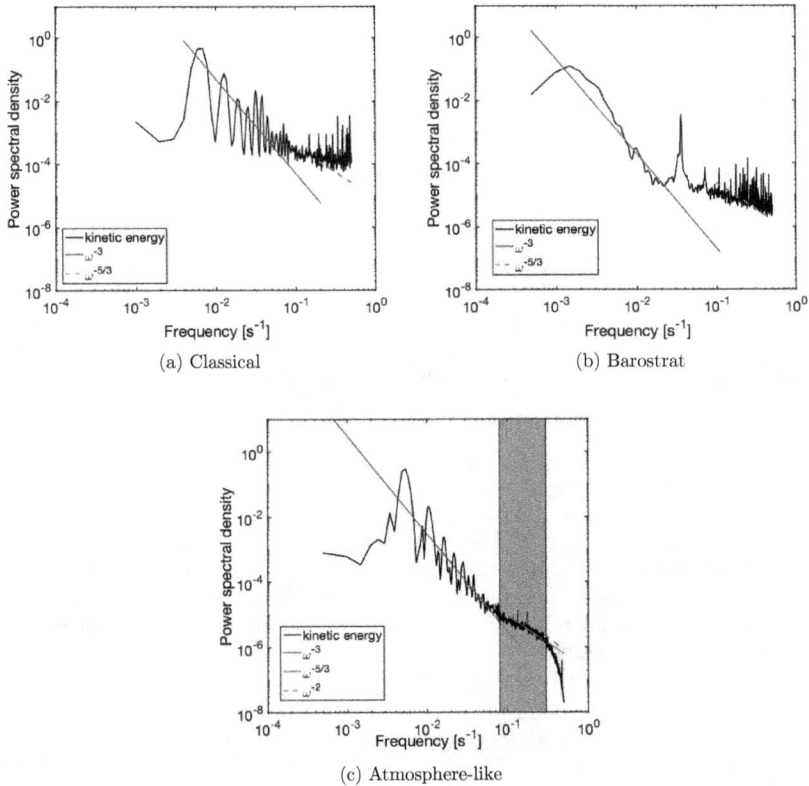

Figure 7.19: Global kinetic energy spectra for different experimental set-ups (a) classical annulus (b) barostrat, and (c) atmosphere-like annulus. The blue and red lines show the slopes ω^{-3} and $\omega^{-5/3}$ respectively. The dashed red line in figure (c) indicates the slope ω^{-2}, for a comparison with the typical trend observed in the oceanic spectra. The region in grey in plot (c) highlights the Doppler shifted gravity wave frequency band.

have large fluctuations, indicating that some noise that could be affecting the data for small frequencies.

Finally, the spectrum calculated for the big-tank (figure 7.19(c)) shows a striking resemblance with the atmospheric spectra in figure 7.18. Although the large scales are not fully resolved, as it can be seen from the large fluctuations at low frequencies, a clear change of slope happens at frequencies around 10^{-1}. For the flatter part, two slopes are shown: $-5/3$ and -2. The latter is the slope based upon the theory by the Garret-Munk (Munk 1981) for internal gravity waves in the ocean. We have mentioned the Garret-Munk model because the internal gravity model for the atmosphere that we use is, in fact, and an extension of the well established oceanic model. VanZandt (1982) showed that for the atmospheric spectra the slope should be changed from -2 to $-5/3$. The two slopes are not very far from one another, as it can be seen in figure 7.19(c), where $-5/3$ fits the most data almost perfectly, suggesting that the experiment has more features in common with the atmosphere than with the oceans in the investigated regime. Another aspect that can be noticed, is the steepening of the spectrum for the highest frequencies. This shows that the measurements resolve the small scales up to dissipation. In figure 7.19(c) we marked the range of Doppler shifted IGW frequencies by the grey colour. Such frequencies are taken from the wavenumber-frequencies plot previously shown in figure 7.12(b) and it can easily be seen that the flatter subrange of the spectrum lies in the measured IGWs frequency range. This already hints that gravity waves might play an important role in terms of energy contribution to the flow at those scales.

The comparison we have just seen is a further element that confirms what is sustained throughout this thesis, i.e. that the classical configuration of the experiment is not suitable to investigate the emission of gravity waves from jets and fronts. Indeed, the fact that the spectra do not show the two different dynamical regimes supports the thesis that the experiment is a good tool to investigate large-scale features, but when it comes to smaller-scale gravity waves it is not a suitable choice. This is further confirmed by the fact that no signal of frontal related gravity waves has ever been observed in the classical continuously stratified annulus.

Even though we have not investigated other regions of the regime diagram and therefore cannot yet confirm the universality of these spectra, we can say from the analysed regimes that the other two configurations are providing a better set-up, both showing features similar to those found in the atmosphere. Some differences in the spectra can be observed when the experimental parameters are modified. In particular, a dependency on the rotation rate has been recently observed by Read et al. (2018) in numerical simulations performed with a global circulation model.

In the following, we focus on the big-tank only and further investigate possible energy exchanges between the different scales, which are believed to be behind the manifestations of such slopes. The wave vortex decomposition method, consisting of a two-step decomposition, is now applied (Bühler et al. 2014). The first step is a Helmholtz decomposition that can be applied to the horizontal velocity spectra to separate the rotational and the divergent component. Figure 7.20(a) and (c) shows the Helmholtz decomposition for the atmospheric (a) and experimental (c) energy spectra, obtained following the procedure explained in 5.2.3.2. The black line shows the kinetic energy

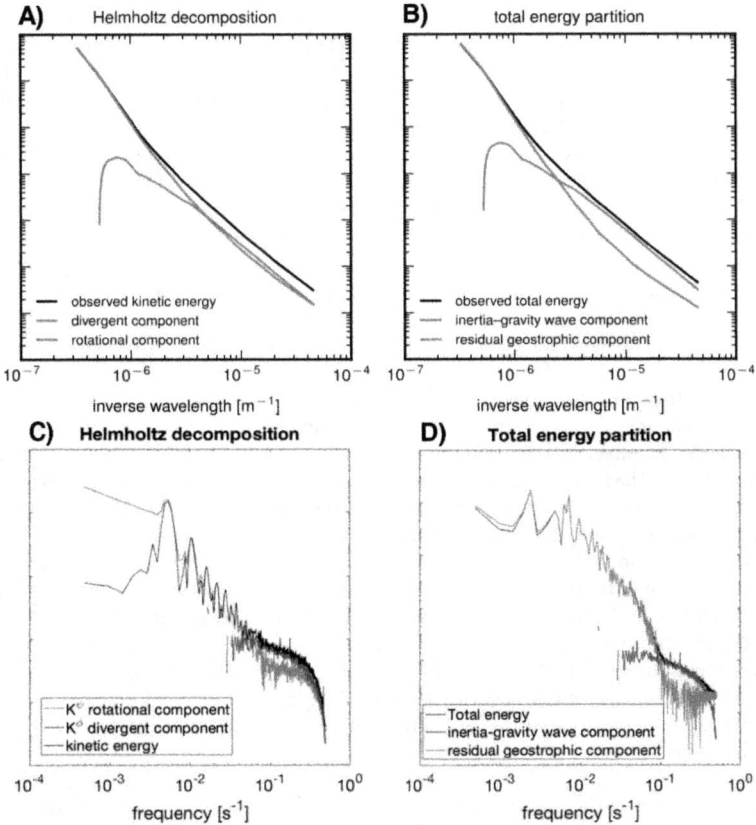

Figure 7.20: Helmholz decomposition of the energy spectra (a) for atmospheric data by Callies et al. (2014) and (c) laboratory experiment. The blue line is the divergent component K^ϕ and the red line is the rotational component K^ψ. Total energy partition spectra (b) for atmospheric data by Callies et al. (2014) and (d) laboratory experiment. The black line is the total energy, the blue line is the energy associated with the inertia-gravity waves, and the red line is the residual energy associated with the geostrophic motions.

(7.10), the red line shows the rotational component K^ψ (5.24a), and the blue line shows
the divergent component K^ϕ (5.24b). For the experiment, the blue and the red curves
intersect at a frequency equal to 4×10^{-2}, which is the same at which the slope flattens
in figure 7.19(c). A feature visible in figure 7.20(c) and later in figure 7.20(d) is that
the divergent energy spectrum, attributed to the subsequently calculated wave energy
spectrum, is truncated at small frequencies. This is because we assumed that our
velocities are isotropic. Bühler et al. (2017) recently showed that, if the data are actually
anisotropic, the divergent energy spectrum becomes negative at small wavenumbers
and extended the method to be valid for anisotropic data as well (see section 5.2.3.2).
This last method involves the further calculation of the cross-spectrum, and under some
assumptions can extend the divergence spectrum to lower energies. However, we did not
apply this method to our data, since the cross-correlation velocity spectrum is difficult
to obtain. The rotational component (red line in figure 7.20) is the one contributing
the most to the energy for small frequencies, whilst the divergent component (blue line)
becomes more important at frequencies in the range of the gravity waves. If we translate
frequencies into wavenumbers, this plot clearly shows that the divergent component
matters at small scales. The experimental spectra (figure 7.20(c)) differ from what was
observed by Callies et al. (2014) and reported in figure 7.20(c). As it can be seen in
the latter, rotational and divergent component of the spectra are comparable at the
mesoscale, although for gravity waves to be the dominant process at the mesoscales one
would expect the divergent component to show higher values. The motivation given
in their paper is that when the inertia-gravity wave field is dominated by near-inertial
waves the two components are expected to be of the same order. This is in opposition to
the findings of Lindborg (2007) who argued that if the two components are at the same
orders the mesoscale motions are not dominated by internal gravity waves and, therefore,
supported the hypothesis of stratified turbulence associated with a forward energy
cascade. Our experiments, on the other hand, support the gravity waves hypothesis.
Since in our experiment $N/f \approx 2$, the range of frequencies for gravity waves is much
narrower than the one in the atmosphere where $N/f \approx 100$. We speculate that this
might lead to the differences between atmospheric aircraft data and the lab data since
in the atmosphere the frequencies are spread more equally from the near-inertial to the
buoyancy range.

The second step is the decomposition of the total energy spectrum into the geostrophic
and inertia-gravity waves component. With respect to the Helmholtz decomposition,
this further step accounts for inertia-gravity waves made of both a divergence and a
rotational component and it is based upon the linear dynamic of inertia-gravity waves.
For calculating the total energy spectrum, we need to consider the potential energy in
addition to the already calculated kinetic energy. The potential energy is calculated from
the temperature measured by the sensors placed at mid gap as

$$E_{\text{pot}} = \frac{1}{2} g^2 \frac{T'^2}{N^2}, \tag{7.11}$$

where T'^2 are the temperature fluctuations relative to the background mean temperature and N is the mean buoyancy frequency. Since the temperatures are measured over
the entire duration of the experiment, the time series have a duration of several hours
and, therefore, resolve the slow frequency of the baroclinic waves. On the other hand,
the temperature spectra appear more disturbed by noise at the high frequencies and a

Gaussian filter has been applied to smooth the plot. This is visible in figure 7.20(d) from the flat tail in the total energy spectrum (in black) which is likely due to the potential energy component calculated from the temperature data that resolve rather well the low frequency and therefore the synoptic baroclinic scale, but on the contrary does not appear to resolve the dissipation scales, as instead we can clearly see from the PIV measurements. Nevertheless, the total energy decomposition plot further shows that the high frequencies are dominated by nearly linear gravity waves, whose energy is comparable to the total energy spectrum. The comparison of figure 7.20(c) and 7.20(d) showing similar energy parts for the divergence and the inertia-gravity wave component also indicates that here the horizontal component is the most energetic part of the gravity waves.

Chapter 8

Conclusions

In this work, we presented the investigation of inertia-gravity wave (IGW) emission from jets and fronts via laboratory experiments. IGWs are known to play an important role in the atmospheric dynamics and represent a significant source of uncertainty in the climate prediction models as they are spatially too small to be fully resolved in such models and their current parametrisation is often oversimplified. One of the least developed is the parametrisation for gravity waves emitted from large-scale baroclinic flows. In this context, numerical simulations and laboratory experiments provide powerful testbeds to validate new theories and parametrisations, as they offer a fully controlled environment and repeatable conditions to study the gravity wave emission from jets.

The experimental investigation presented in this thesis has been done using a differentially heated rotating annulus experiment, which was chosen because of its proven capability to capture the fundamental physics of atmospheric dynamics (Hide 1958, Von Larcher and Egbers 2005, Read et al. 2014, Vincze et al. 2014). A long-standing problem is whether instabilities of baroclinic fronts, and in particular spontaneous imbalance of the frontal flow can be observed in differentially rotating annulus experiments. Short-wave patterns have been observed in shear-driven rotating experiments by Lovegrove et al. (2000) and Williams et al. (2005) where spontaneous emission has been invoked as the generating mechanism. However, other studies (Flór et al. 2011) have underlined the role played by other mechanisms, and therefore the source for these patterns is not fully clear yet. The only experimental observations of IGWs in the differentially heated rotating annulus known to the author are those reported by Read (1992), followed by the study by Jacoby et al. (2011), who identified the source of such gravity waves in the inner wall boundary layer instability. One open question left from the studies by Jacoby et al. (2011) is whether other source types, in particular spontaneous emission, could also be studied in the differentially heated rotating annulus.

What emerges from this thesis is that, in fact, the baroclinic rotating annulus in its classical configuration is not ideal for investigating emission and propagation of atmospheric IGWs from jets and fronts.

A detailed investigation of the problems related to the classical baroclinic annulus set-up has been presented in chapter 3, where two alternative set-ups are also introduced. The core of the problem is that to have baroclinic waves generating atmosphere-like IGWs, a small inverse aspect ratio $\Pi = H/L$, where H is the fluid depth and L the gap width, and a large N/f, with N the buoyancy frequency and f the Coriolis frequency, are required. However, as we have experimentally proven, for annuli having a gap width of the order of centimetres (as experiments in the classical configuration usually do), it is impossible to reach a low value of the inverse aspect ratio without having to deal with the viscous Ekman layer, which inhibits the formation of a steady baroclinic wave. Therefore, we proposed two alternative and more suitable set-ups, namely the barostrat and the atmosphere-like experiments, which both lead to the observation of inertia-gravity waves along the baroclinic jet.

The barostrat experiment is a thermohaline version of the classical differentially heated rotating annulus set-up. In the barostrat experiment, the geometry of the rotating tank remains the same as in the classical experiments by Hide and others, but instead of strat-ification due to differentially heating the lateral walls alone, a salt stratification is added in the axial direction. The introduction of salt stratification has two main effects: the first is that N is directly increased, and the second is that thin convectively mixed layers form at the top and the bottom of the tank where baroclinic instability can take place and a thicker stable stratified dynamically less active layer arises in between. Therefore, the two conditions we postulated necessary for atmosphere-like gravity waves and baroclinic waves to occur are satisfied.

The second solution proposed by Borchert et al. (2014) is a newly built tank, with a much larger horizontal diameter. By increasing the horizontal dimensions of the experi-ment, we obtained an aspect ratio equal to half the one of the classical configuration for a fluid depth $z = 6\,\text{cm}$, which granted the formation of baroclinic waves undisturbed by the bottom viscous Ekman layer. Moreover, by decreasing the aspect ratio, the baroclinic instability regime is reached for higher values of N/f which, as we have seen, is a crucial point for the generation and propagation of IGWs. Indeed, a larger N/f allows for a broader band of intrinsic frequencies for gravity waves, meaning that we can expect to observe a broader spectrum of propagating and trapped waves in the experiment.

The dynamics of the large-scale baroclinic waves and small-scale gravity waves have been investigated in detail and reported in chapter 6 for the barostrat experiment and chapter 7 for the atmosphere-like experiment. In the following, the main results are summarised.

8.1   Barostrat experiment

In chapter 6, we have experimentally investigated and demonstrated that the thermohaline version of the differentially heated rotating annulus is not only a testbed to study large-scale wave interactions but is also a set-up suitable to study interactions of different wave types, including IGWs. The introduction of a stable salinity profile leads to the double-diffusive convection and, consequently, to the natural formation of layers which offers the possibility to investigate wave resonance but also wave propagation between the different layers.

For the parameters investigated, two separate baroclinic waves can be observed forming in two unstable convective cells confined at the top and the bottom of the tank. The two baroclinic waves show different features having azimuthal wave numbers $m = 3$ for the top and $m = 4$ for the bottom layer. The baroclinic wave at the top is in a steady state with a slow drift rate and excites higher harmonics by self-interaction. The baroclinic wave at the bottom shows amplitude vacillations and a fast drift. The vacillations result from the interaction between two $m = 4$ waves with different phase speeds, as it has been described by Buzyna et al. (1989) for a classical annulus experiment. Though nonlinear triadic interaction between the waves in the individual layers can clearly be seen in the spectra, it is not yet clear whether the baroclinic waves in the surface and bottom layers are coupled. What is striking is the close correspondence of some prominent peaks in the spectra: the steady wave peak in the top layer is very close to the difference of the frequencies of the two dominant wavenumber 4 modes in the bottom layer implying a 'frequency triad' between the top and bottom layer waves. Moreover, the two harmonic frequencies of the upper layer wave nearly correspond with the frequency of the two wavenumber 4 waves in the bottom layer.

Besides the large-scale baroclinic waves, we further observe an inertial Kelvin type global mode and a higher frequency surface wave of Poincaré type. The inertial Kelvin mode shows similarity with a shallow water boundary trapped Kelvin wave, has an azimuthal wavenumber 1, but has the frequency of the tank's rotation and not f. It is, hence, not a shallow water mode but an inertial wave Kelvin mode with a certain vertical structure. Unfortunately, since we did not measure simultaneously at different vertical levels we do not know the vertical wavenumber. We just know that the inertial Kelvin mode is strongest in the non-convective zone and the bottom layer. The mode is very likely mechanically driven by a slight deviation of the rotation axis from the vertical.

The origin of the Poincaré type modes with frequencies larger than f is not yet clear. They might also be triggered mechanically by a very weak sloshing of the tank. A theoretical treatment of those modes (as done by Mougel et al. (2015) for the homogeneous case) is hampered by the nonlinear vertical density profile. For a thorough analysis effects at the internal interfaces have to be considered in addition to the surface wave modes. Such a study is postponed to the future for when more data, in particular in the high frequency range and from vertical cross sections, will be available.

In contrast to the global modes described above, short-wave inertia-gravity waves are hard to be found in the wave spectra since they typically occur sporadically and are spatially localised as they are tightly connected to the baroclinic front. Although spontaneous imbalance has been documented from a number of numerical simulations also for the annulus configuration (Borchert et al. 2014, Hien et al. 2018) for which the Reynolds number, like in the described experiment, was rather small ($Re \simeq 100$), it has never been experimentally observed for the continuously stratified annulus. For the latter, in general, the Rossby numbers are rather small and the flow is in a nearly balanced state. In our experiments the top layer shows wave packets in the horizontal divergence field travelling mainly with the frontal mean flow and hence much faster than the drift speed of the baroclinic waves. These structures are not unlike the wave trains described by O'Sullivan and Dunkerton (1995) and we have shown that their signal is above the noise level, their frequencies are in the inertia-gravity wave range and their corresponding frequency-wavenumber plot roughly follows one branch of the dispersion curves. These observations support what was said before about the thermohaline version of the experiment being more suitable to study gravity waves since the ratio N/f is, in general, larger than one in contrast to the classical differentially heated rotating annulus for which the ratio is much smaller than one. As in the atmosphere, gravity is playing a more prominent role for the frontal inertia-gravity waves in the thermohaline experiment. For the classical annulus, inertia is dominant and the baroclinic fronts do occur over the full depth of the tank and not just in thin layers. The latter might also be more favourable to generate a stronger jet flow and hence frontal waves.

8.2   Atmosphere-like tank

The new atmosphere-like tank, especially built and tested at the BTU laboratories to study IGWs, is presented for the first time in this thesis. This wider and shallower configuration follows theoretical considerations and was proposed for the first time in the numerical study by Borchert et al. (2014). The finite volume model, firstly used by Borchert et al. (2014) and then by Hien et al. (2018), allows to run numerical simulations having the same geometry of the experiment and directly correspondent lateral temperature boundary conditions and rotation rates can be applied. With this model (although for a lateral temperature difference higher than the maximum reachable with the laboratory experiment), Hien et al. (2018) proved that IGWs can be generated not only by the boundary instabilities at the lateral walls, but also by spontaneous imbalance of the baroclinic jet.

The first part (sections 7.1 to 7.1.4) of chapter 7 was devoted to the investigation of large-scale baroclinic waves and the consequences such flow can have on the propagation of gravity waves. More specifically, we investigated in detail the vertical and radial temperature fields, comparing data recorded in the laboratory experiment with the numerical simulations. In general, even though some discrepancies can be expected because of the approximation made in the numerical simulations of an adiabatic and wind stress-free surface, a good quantitative and qualitative match has been observed between the two. What we observed in the experiment is that the two temperature gradients along the vertical and horizontal direction do not match. Hide (1967) introduced the factors σ_z and σ_r as a measure of the vertical and horizontal temperature gradient respectively and found that $\sigma_z > \sigma_r$ for the classical configuration. Our studies show instead that this ratio reverses, i.e. $\sigma_z < \sigma_z$ for the atmosphere-like configuration. Since $Bu = \sigma_z Ro_T$, it follows that $Fr = (Ro_T/\sigma_z)^{(1/2)}$ and because σ_z increases for geostrophic turbulent flows, it suggests that for this flow regime the background conditions come closer to the one considered for Lighthill radiation since Fr becomes smaller as required. Note that still the wave source, considered to be small compared to the long gravity wave in shallow water, is different for the stratified case. For this reason, caution should be exercised when bringing spontaneous emission in stratified fluids in the context of Lighthill radiation that strictly holds only for shallow water flows.

The detailed study of horizontal and vertical maps of the buoyancy frequency has shown how atmosphere-like IGWs can, in principle, be generated and propagate over almost the entire fluid domain. However, because of the larger variations observed in the laboratory experiment compared to the numerical simulations, we speculate that there are regions that are forbidden for the gravity waves to propagate. This is confirmed by the observation of waves captured along the baroclinic jet, where the buoyancy frequency is observed to reach its maximum value. Plougonven and Snyder (2005) showed with numerical simulations how the wave capture mechanism might be useful to predict the location and several characteristics of spontaneously generated gravity waves.

The second part (from section 7.2) of chapter 7 focussed on the study of the small-scale wave features. All the experiments investigated show consistent small wave activity along the jet front, particularly at the entrance and exit regions, which are known to be hot spots where gravity waves can be observed in the atmosphere and numerical simulations (O'Sullivan and Dunkerton 1995, Plougonven and Snyder 2005, von Storch et al. 2019). We have shown how such waves are continuously emitted along the baroclinic jet and how they propagate along with it. These waves have characteristics similar to the waves observed in the shear-driven rotating annulus experiment by Lovegrove et al. (2000) and in the barostrat experiment (Rodda et al. 2018). A two dimensional Fourier analysis led to the identification of frequencies and wavenumbers and the location of wave energy maxima. The frequencies and wavenumbers satisfy the Doppler shifted dispersion relation for gravity-waves, indicating that the observed small-scale features are, indeed, IGWs.

One of the most important steps towards the parametrisation of gravity waves emitted from jets and fronts is the investigation of the relation between the IGW amplitude and the Rossby number. Several theories predict exponentially small amplitude for IGWs generated by balanced motions (e.g. Vanneste and Yavneh (2004)), whilst others suggest algebraically small amplitudes (Pedlosky 1982). The experimental results reported by Williams et al. (2008) suggest that IGW amplitude varies linearly with the Rossby number. Although our results on the gravity wave scaling are only preliminary and more data are needed to correctly quantify the scaling, we can tentatively say that gravity wave activity increases for larger Rossby numbers, in agreement with existing theories. These first results are promising and show how the atmosphere-like experiment could help in developing new theories and be used for testing existing parametrisations.

Another aspect we have investigated is the generation mechanism responsible for the gravity wave emission. At the baroclinic wave front, several phenomena can occur and trigger gravity waves. We identify the main mechanisms as Kelvin-Helmholtz instability, convection, and spontaneous imbalance. In addition to these mechanisms located at the fronts, boundary wall instability can also generate gravity waves. Our investigation shows that Kelvin-Helmholtz instability and boundary wall instability can be ruled out. Convection and spontaneous emission, instead, are both possible. Furthermore, we have seen that at the baroclinic exit region gravity waves can be influenced by another mechanism, namely the wave capture, which determine the properties of the waves as they propagate through regions with strong flow deformation. Since the occurrence of IGWs is strongly coupled with an enhancement of the local Rossby number and we could rule out convection in experiments with larger temperature difference, we speculate that the most likely generation mechanism is spontaneous emission.

A long standing open question concerns the mesoscale portion of the atmospheric energy spectra. Several different theories about the phenomena involved have been proposed but the source of energy at these small-scales is not clear yet. The energy spectra obtained from the atmosphere-like experiment are in exceptionally good agreement with the atmospheric spectra by Callies et al. (2014), further validating the relevance of such experiment for atmospheric studies on multi-scale interactions. By applying the method developed by Bühler et al. (2014) to our experimental data for analysing the atmospheric spectra, we could attribute most of the energy at the mesoscale to inertia-gravity waves.

Our results, therefore, support the analysis byCallies et al. (2014) who associated the energy at the mesoscale with inertia-gravity waves.

8.3   Open questions and future work

On the base of experimental data alone, not all questions related to IGW generation mechanisms in baroclinic jets can be answered. For this reason, there remains some space for speculation as the reader has noticed. Therefore, more numerical simulations are necessary to complement the experimental data. For the barostrat experiment no stable numerical model is available yet. In contrast, for the atmosphere-like tank a corresponding numerical model exists. However, one of the main open questions emerging from this thesis concerns the differences between the inertia-gravity waves observed in the laboratory and in the numerical simulations. Despite the good agreement obtained in terms of the large-scale wave regime, we have identified some discrepancies in the temperature and, consequently, in the buoyancy frequency field. We attribute such discrepancies to the different upper boundary condition used for the numerical simulations, which is adiabatic and does not take into account heat exchange between the water surface and the air above it. To address this, a lid for the experiment has been built and tested whilst test cases with a modified version of cyFloit, including an upper boundary condition which allows for the heat exchange (based on the work by Diekmann (2014)) are currently running. The output of the simulations will then be analysed and compared to the experimental data to get a clearer picture of the effect of the upper boundary on gravity waves in the annulus.

Another aspect we investigated is the gravity wave amplitude dependency on the Rossby number. The results found are preliminary and further investigation, possibly spanning over a larger range of Ro are needed for more conclusive validation. Furthermore, so far we have focused on gravity wave emission from baroclinic waves in steady regime. Investigations of other flow regimes, such as vacillating unsteady flow, are relevant to determine what the background flow conditions are that enhance gravity wave emission. Experiments exploring unsteady flow regimes are planned for the near future as a continuation of the work presented in this thesis. The experimental findings will provide a consistent support to the development of a suitable parametrisation for spontaneously emitted gravity waves in the climate models.

Appendices

A.1 Calculation of the determinant of the matrix

Consider the matrix

$$
A = \begin{bmatrix}
\rho_0 \partial_t & 0 & 0 & 0 & \partial_x \\
0 & \rho_0 \partial_t & 0 & 0 & \partial_y \\
0 & 0 & \rho_0 \partial_t & g & \partial_z \\
\partial_x & \partial_y & \partial_z & 0 & 0 \\
0 & 0 & -N_0^2 \rho_0/g & \partial_t & 0
\end{bmatrix}
\tag{A.1}
$$

Since the matrix has constant coefficients and therefore the differentiation is commutative we can take the determinant

$$
det(A) = 0.
\tag{A.2}
$$

That can be calculated explicitly via the Leibniz formula in the following way

$$
|A| = \begin{vmatrix}
\rho_0 \partial_t & 0 & 0 & 0 & \partial_x \\
0 & \rho_0 \partial_t & 0 & 0 & \partial_y \\
0 & 0 & \rho_0 \partial_t & g & \partial_z \\
\partial_x & \partial_y & \partial_z & 0 & 0 \\
0 & 0 & -N_0^2 \rho_0/g & \partial_t & 0
\end{vmatrix}
$$

$$
= \rho_0 \partial_t
\begin{vmatrix}
\rho_0 \partial_t & 0 & 0 & \partial_y \\
0 & \rho_0 \partial_t & g & \partial_z \\
\partial_y & \partial_z & 0 & 0 \\
0 & -N_0^2 \rho_0/g & \partial_t & 0
\end{vmatrix}
+ \partial_x
\begin{vmatrix}
0 & \rho_0 \partial_t & 0 & 0 \\
0 & 0 & \rho_0 \partial_t & g \\
\partial_x & \partial_y & \partial_z & 0 \\
0 & 0 & -N_0^2 \rho_0/g & \partial_t
\end{vmatrix}
$$

$$
= \rho_0 \partial_t \left(\rho_0 \partial_t
\begin{vmatrix}
\rho_0 \partial_t & g & \partial_z \\
\partial_z & 0 & 0 \\
-N_0^2 \rho_0/g & \partial_t & 0
\end{vmatrix}
- \partial_y
\begin{vmatrix}
0 & \rho_0 \partial_t & g \\
\partial_y & \partial_z & 0 \\
0 & -N_0^2 \rho_0/g & \partial_t
\end{vmatrix}
\right) +
$$

$$
+ \partial_x \left(-\rho_0 \partial_t
\begin{vmatrix}
0 & \rho_0 \partial_t & g \\
\partial_x & \partial_z & 0 \\
0 & -N_0^2 \rho_0/g & \partial_t
\end{vmatrix}
\right)
$$

Solving the last determinants we get

$$\rho_0^2 \partial_t^3 \partial_z^2 + \rho_0^2 \partial_t^3 \partial_y^2 + \rho_0^2 N_0^2 \partial_t \partial_y^2 + \rho_0^2 \partial_t^3 \partial_x^2 + \rho_0^2 N_0^2 \partial_t \partial_x^2 = 0.$$

Simplifying and re-arranging the terms of the equation we get

$$\partial_t^2 (\partial_x^2 + \partial_y^2 + \partial_z^2) + N_0^2 (\partial_x^2 + \partial_y^2) = 0. \tag{A.4}$$

By definition

$$\nabla^2 = \partial_x^2 + \partial_y^2 + \partial_z^2, \qquad \nabla_H^2 = \partial_x^2 + \partial_y^2, \tag{A.5}$$

therefore the final expression is

$$\partial_t^2 \nabla^2 + N_0^2 \nabla_H^2 = 0. \tag{A.6}$$

A.2 Dispersion relation

We consider equation

$$\frac{\partial^2}{\partial t^2}\left(\nabla^2 w\right) + N^2 \nabla_h^2 w = 0. \tag{A.7}$$

where

$$\nabla^2 = \frac{\partial^2}{\partial x^2} + \frac{\partial^2}{\partial y^2} + \frac{\partial^2}{\partial z^2}, \qquad \nabla_H^2 = \frac{\partial^2}{\partial x^2} + \frac{\partial^2}{\partial y^2}. \tag{A.8}$$

We want to solve it assuming solution of the form

$$w = W_0 \exp[i(kx + ly + mz - \omega t)]. \tag{A.9}$$

We calculate the first derivatives

$$\frac{\partial w}{\partial x} = ikw, \quad \frac{\partial w}{\partial y} = ilw, \quad \frac{\partial w}{\partial z} = imw, \quad \frac{\partial w}{\partial t} = -i\omega w, \tag{A.10}$$

and the second derivatives

$$\frac{\partial^2 w}{\partial x^2} = -k^2 w, \quad \frac{\partial^2 w}{\partial y^2} = -l^2 w, \quad \frac{\partial^2 w}{\partial z^2} = -m^2 w, \quad \frac{\partial^2 w}{\partial t^2} = -\omega^2 w, \tag{A.11}$$

Substituting into (A.7) we have

$$\frac{\partial^2}{\partial t^2}\left(-(k^2 + l^2 + m^2)\right) w + N^2(-(k^2 + l^2))w = 0 \tag{A.12}$$

and then

$$-\left(k^2 + l^2 + m^2\right)(-\omega^2)w - N^2(k^2 + l^2)w = 0. \tag{A.13}$$

The dispersion relation is

$$\omega^2 = \frac{N^2(k^2 + l^2)}{(k^2 + l^2 + m^2)} = \frac{N^2 K_H^2}{K^2} \tag{A.14}$$

A.3   Ray equations

The ray theory works for almost planar waves, which locally can be approximated as planar wave but change in space with the assumption that $\lambda << L$, i.e. the variations are slow in space. Most of this appendix is taken from the lecture notes by Harlander (2014).

The wave phase is

$$\Theta(\vec{x}, t) = \vec{k} \cdot \vec{x} - \omega t. \tag{A.15}$$

We can write the local wavenumber vector as

$$\vec{k}(\vec{x}, t) = \nabla\Theta, \tag{A.16}$$

and local frequencies as

$$\omega(\vec{x}, t) = -\frac{\partial \Theta}{\partial t}. \tag{A.17}$$

By taking the gradient of (A.17)

$$\nabla \cdot \left(\frac{\partial \Theta}{\partial t} + \omega(\vec{x}, t) \right) = \frac{\partial \vec{k}}{\partial t} + \nabla\omega(\vec{x}, t) = 0, \tag{A.18}$$

which is called the 'eikonal equation'. The dispersion relation can now be written as

$$\omega(\vec{x}, t) = \Omega(\vec{k}(\vec{x}, t), \vec{x}, t), \tag{A.19}$$

where Ω has an explicit dependency on $\vec{x}$ and t, but also an implicit dependency as $\vec{k}$ depends on $\vec{x}$ and t itself. Taking the time derivative, we have

$$\frac{\partial \omega}{\partial t} = \frac{\partial \Omega}{\partial t} + \frac{\partial \Omega}{\partial k}\frac{\partial k}{\partial t} + \frac{\partial \Omega}{\partial l}\frac{\partial l}{\partial t} + \frac{\partial \Omega}{\partial m}\frac{\partial m}{\partial t}. \tag{A.20}$$

If we use the definition of group velocity $\vec{c}_g = (\partial\Omega/\partial k, \partial\Omega/\partial l, \partial\Omega/\partial m)$, we can rewrite

$$\frac{\partial \omega}{\partial t} = \frac{\partial \Omega}{\partial t} + \vec{c}_g \frac{\partial \vec{k}}{\partial t}. \tag{A.21}$$

Using (A.18) and defining $d_g/dt = \partial/\partial t + \vec{c}_g \cdot \nabla$, we obtain

$$\frac{d_g\omega}{dt} = \frac{\partial \Omega}{\partial t}. \tag{A.22}$$

We can proceed in a similar way to get an equation for the wave number. This time we have the spatial derivatives

$$\frac{\partial \omega}{\partial x} = \frac{\partial \Omega}{\partial x} + \frac{\partial \Omega}{\partial k}\frac{\partial k}{\partial x} + \frac{\partial \Omega}{\partial l}\frac{\partial l}{\partial x} + \frac{\partial \Omega}{\partial m}\frac{\partial m}{\partial x} \tag{A.23a}$$

$$\frac{\partial \omega}{\partial y} = \frac{\partial \Omega}{\partial y} + \frac{\partial \Omega}{\partial k}\frac{\partial k}{\partial y} + \frac{\partial \Omega}{\partial l}\frac{\partial l}{\partial y} + \frac{\partial \Omega}{\partial m}\frac{\partial m}{\partial y} \tag{A.23b}$$

$$\frac{\partial \omega}{\partial z} = \frac{\partial \Omega}{\partial z} + \frac{\partial \Omega}{\partial k}\frac{\partial k}{\partial z} + \frac{\partial \Omega}{\partial l}\frac{\partial l}{\partial z} + \frac{\partial \Omega}{\partial m}\frac{\partial m}{\partial z}. \tag{A.23c}$$

The set of equation can be written in a compact form

$$\nabla \omega = \nabla \Omega + \vec{c}_g \nabla \vec{k}. \tag{A.24}$$

Using the eikonal equation to get

$$\frac{d_g \vec{k}}{dt} = -\nabla \Omega. \tag{A.25}$$

Altogether, the ray tracing equations are

$$\frac{d_g \vec{k}}{dt} = -\nabla \Omega \tag{A.26a}$$

$$\frac{d_g \omega}{dt} = \frac{\partial \Omega}{\partial t} \tag{A.26b}$$

$$\frac{d_g \vec{x}}{dt} = \vec{c}_g \tag{A.26c}$$

This set of equations are satisfied by wave packets propagating along paths (also called rays) that are time dependent.

A.4   Following the wave

We show here the extension to the entire time series of the analysis done in section 7.2.2. When considering a long time series, because of the baroclinic wave moving through the domain over time, also the shape and the length of the streamlines will change. An example of this can be seen in figure A.1 (a) and (b), where two snapshots (at $t_0 = 1\,\text{s}$ and $t_1 = 30\,\text{s}$) of the horizontal divergence and the chosen streamlines are shown. To better visualise the variations of the streamlines and the consequences this has on the orientation of the gravity waves, the along-the-streamline–time plot is shown in figure A.1 (c). The red and black circles indicate the radial position (visible on the y-axis placed on the right of the plot) of the start and end point of the streamlines. These circles are indicated both in figure A.1(a) and (b). When the two circles close up, the streamline length increases. The complete plot obtained for the full-time series is visible in figure A.1(d). The darker blue areas at the bottom of the plot are generated by the last points belonging to the streamline falling in the regions close to the edges of the domain, where the divergence data are rather uncertain.

The final step before computing the Fourier transform is to cut the data and consider only the central regions (which correspond to the shortest length of the streamline) so that we have the same length for all of them and we eliminate the dark blue regions at the bottom. The resulting plot is shown in figure A.2(a), where the enhancement of the divergence signal in correspondence of the three baroclinic wave jets is even more prominent than in the previous Hovmöller plots.

(a) streamline at $t = 1$ s

(b) streamline at $t = 30$ s

(c)

(d)

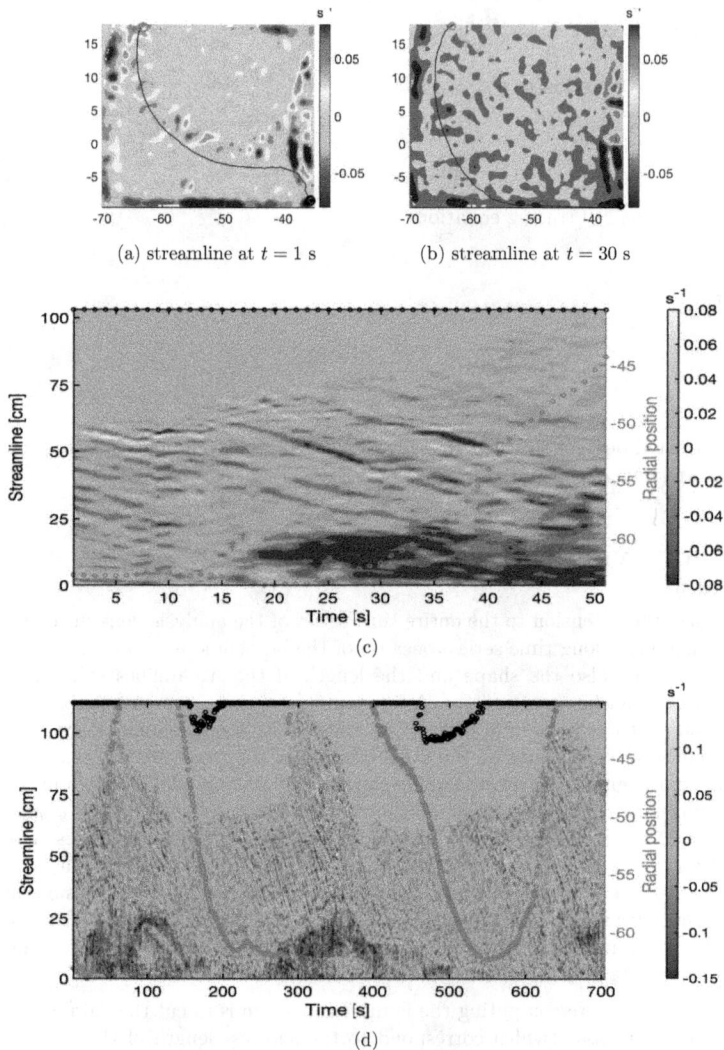

Figure A.1: (a) and (b) show two examples of the streamlines chosen to analyse the horizontal divergence along with the red and black circles indicating the starting and the ending points of the streamlines, (c) Hovmöller plot of the horizontal divergence for $1s \leq t \leq 50s$, and (d) Hovmöller plot of the horizontal divergence for $1s \leq t \leq 700s$. The red and black markers in (c) and (d) indicate the position along the radial direction of the beginning and the end of the streamline at every time step considered.

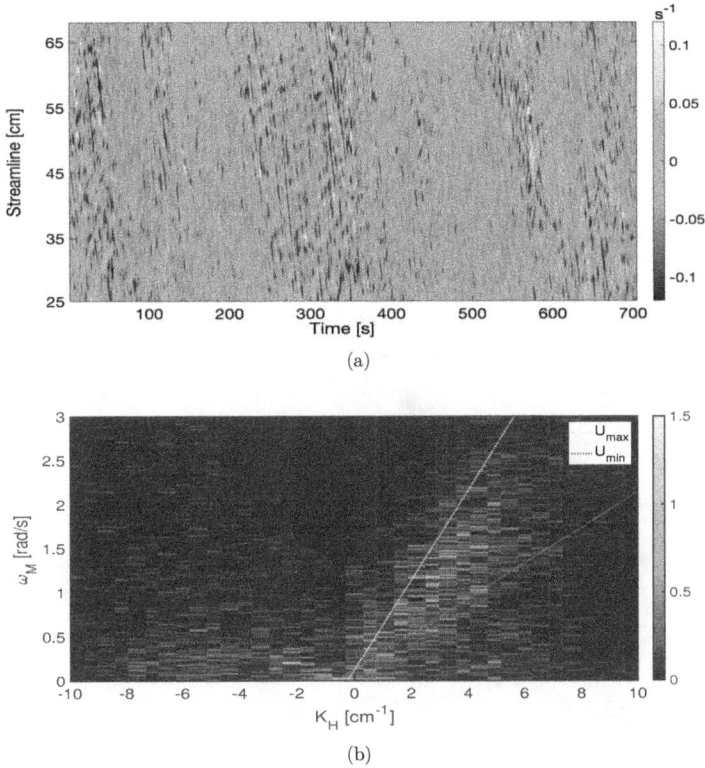

Figure A.2: (a) Hovmöller plot of the horizontal divergence for $1s \leq t \leq 700s$ (as in figure A.1(d), but the data at the top and the bottom of the plot are cut out). (b) 2D fft plot with dispersion curves for IGWs calculated with (7.6). The curve in white is doppler shifted using the maximum velocity $U_{\max}$ measured along the jet and the red one with the minimum velocity $U_{\min}$.

References

M. Abramowitz and I. A. Stegun. *Handbook of mathematical functions: with formulas, graphs, and mathematical tables*, volume 55. Courier Corporation, 1965.

T. Astarita and G. M. Carlomagno. *Infrared thermography for thermo-fluid-dynamics*. Springer Science & Business Media, 2012.

A. K. Banerjee, A. Bhattacharya, and S. Balasubramanian. Experimental study of rotating convection in the presence of bi-directional thermal gradients with localized heating. *AIP Advances*, 8(11):115324, 2018.

V. Barcilon. Role of the Ekman layers in the stability of the symmetric regime obtained in a rotating annulus. *Journal of the Atmospheric Sciences*, 21(3):291–299, 1964.

D. Benielli and J. Sommeria. Excitation of internal waves and stratified turbelence by parametric instability. *Dynamics of atmospheres and oceans*, 23(1-4):335–343, 1996.

S. Borchert, U. Achatz, and M. D. Fruman. Gravity wave emission in an atmosphere-like configuration of the differentially heated rotating annulus experiment. *Journal of Fluid Mechanics*, 758:287–311, 2014.

O. Bühler and M. E. McIntyre. Wave capture and wave–vortex duality. *Journal of Fluid Mechanics*, 534:67–95, 2005.

O. Bühler, M. E. McIntyre, and J. F. Scinocca. On shear-generated gravity waves that reach the mesosphere. Part I: Wave generation. *Journal of the atmospheric sciences*, 56(21):3749–3763, 1999.

O. Bühler, J. Callies, and R. Ferrari. Wave–vortex decomposition of one-dimensional ship-track data. *Journal of Fluid Mechanics*, 756:1007–1026, 2014.

O. Bühler, M. Kuang, and E. G. Tabak. Anisotropic helmholtz and wave–vortex decomposition of one-dimensional spectra. *Journal of Fluid Mechanics*, 815:361–387, 2017.

G. Buzyna, R. L. Pfeffer, and R. Kung. Kinematic properties of wave amplitude vacillation in a thermally driven rotating fluid. *Journal of the Atmospheric Sciences*, 46(17):2716–2730, 1989.

J. Callies, R. Ferrari, and O. Bühler. Transition from geostrophic turbulence to inertia–gravity waves in the atmospheric energy spectrum. *Proceedings of the National Academy of Sciences*, 111(48):17033–17038, 2014.

J. G. Charney. The dynamics of long waves in a baroclinic westerly current. *Journal of Meteorology*, 4(5):136–162, 1947.

C. Chen, D. Briggs, and R. Wirtz. Stability of thermal convection in a salinity gradient due to lateral heating. *International Journal of Heat and Mass Transfer*, 14(1):57IN163–62IN365, 1971.

J. Y. Cho and E. Lindborg. Horizontal velocity structure functions in the upper troposphere and lower stratosphere: 1. observations. *Journal of Geophysical Research: Atmospheres*, 106 (D10):10223–10232, 2001.

M. Chouksey, C. Eden, and N. Brüggemann. Internal gravity wave emission in different dynamical regimes. *Journal of Physical Oceanography*, 48(8):1709–1730, 2018.

B. Cushman-Roisin and J.-M. Beckers. *Introduction to geophysical fluid dynamics: physical and numerical aspects*, volume 101. Academic press, 2011.

S. B. Dalziel, M. D. Patterson, C. Caulfield, and S. Le Brun. The structure of low-Froude-number lee waves over an isolated obstacle. *Journal of Fluid Mechanics*, 689:3–31, 2011.

C. J. Diekmann. Erweiterung eines numerischen modells des rotierenden annulus um die wärmeübertragung zwischen flüssigkeit und umgebung. Bachelor thesis, 2014.

A. Dörnbrack, M. Pitts, L. Poole, Y. Orsolini, K. Nishii, and H. Nakamura. The 2009-2010 arctic stratospheric winter–general evolution, mountain waves and predictability of an operational weather forecast model. *Atmospheric Chemistry & Physics Discussions*, 11(12), 2011.

A. Dörnbrack, S. Gisinger, N. Kaifler, T. C. Portele, M. Bramberger, M. Rapp, M. Gerding, J. Söder, N. Žagar, and D. Jelić. Gravity waves excited during a minor sudden stratospheric warming. *Atmospheric Chemistry and Physics*, 18(17):12915–12931, 2018.

H. Douglas and P. Mason. Thermal convection in a large rotating fluid annulus: some effects of varying the aspect ratio. *Journal of the Atmospheric Sciences*, 30(6):1124–1134, 1973.

P. Drazin. Variations on a theme of Eady. In *Rotating fluids in geophysics*, chapter 3, pages 139–169. Academic Press, New York, 1978.

D. G. Dritschel and A. Viúdez. A balanced approach to modelling rotating stably stratified geophysical flows. *Journal of Fluid Mechanics*, 488:123–150, 2003.

E. T. Eady. Long waves and cyclone waves. *Tellus*, 1(3):33–52, 1949.

M. Ern, F. Ploeger, P. Preusse, J. Gille, L. Gray, S. Kalisch, M. Mlynczak, J. Russell III, and M. Riese. Interaction of gravity waves with the qbo: A satellite perspective. *Journal of Geophysical Research: Atmospheres*, 119(5):2329–2355, 2014.

M. Ern, P. Preusse, and M. Riese. Driving of the sao by gravity waves as observed from satellite. In *Annales geophysicae*, volume 33, pages 483–504. Copernicus GmbH, 2015.

F. M. Exner. *Über die bildung von Windhosen und Zyklonen*. Hölder-Pichler-Tempsky, 1923.

J. S. Fein. An experimental study of the effects of the upper boundary condition on the thermal convection in a rotating, differentially heated cylindrical annulus of water. *Geophysical & Astrophysical Fluid Dynamics*, 5(1):213–248, 1973.

A. Fincham and G. Spedding. Low cost, high resolution DPIV for measurement of turbulent fluid flow. *Experiments in Fluids*, 23(6):449–462, 1997.

J.-B. Flór, H. Scolan, and J. Gula. Frontal instabilities and waves in a differentially rotating fluid. *Journal of Fluid Mechanics*, 685:532–542, 2011.

R. Ford. Gravity wave radiation from vortex trains in rotating shallow water. *Journal of Fluid Mechanics*, 281:81–118, 1994.

D. C. Fritts and M. J. Alexander. Gravity wave dynamics and effects in the middle atmosphere. *Reviews of geophysics*, 41(1), 2003.

D. C. Fritts and G. D. Nastrom. Sources of mesoscale variability of gravity waves. Part II: Frontal, convective, and jet stream excitation. *Journal of the Atmospheric Sciences*, 49(2): 111–127, 1992.

W.-G. Früh. Amplitude Vacillation in Baroclinic Flows. *Modeling Atmospheric and Oceanic Flows: Insights from Laboratory Experiments and Numerical Simulations*, 205, 2014.

W.-G. Früh and P. Read. Wave interactions and the transition to chaos of baroclinic waves in a thermally driven rotating annulus. *Philosophical Transactions of the Royal Society of London A: Mathematical, Physical and Engineering Sciences*, 355(1722):101–153, 1997.

D. Fultz. Experimental analogies to atmospheric motions. In *Compendium of meteorology*, pages 1235–1248. Springer, 1951.

D. Fultz, R. R. Long, G. V. Owens, W. Bohan, R. Kaylor, and J. Weil. Studies of thermal convection in a rotating cylinder with some implications for large-scale atmospheric motions. *Meteorological Monographs*, 21(4):1–104, 1959.

K. Gage. Evidence far ak- 5/3 law inertial range in mesoscale two-dimensional turbulence. *Journal of the Atmospheric Sciences*, 36(10):1950–1954, 1979.

K. Gage and G. Nastrom. Theoretical interpretation of atmospheric wavenumber spectra of wind and temperature observed by commercial aircraft during gasp. *Journal of the Atmospheric Sciences*, 43(7):729–740, 1986.

M. Ghil, P. Read, and L. Smith. Geophysical flows as dynamical systems: the influence of Hide's experiments. *Astronomy & Geophysics*, 51(4):4–28, 2010.

D. Guimbard, S. Le Dizès, M. Le Bars, P. Le Gal, and S. Leblanc. elliptic instability of a stratified fluid in a rotating cylinder. *Journal of Fluid Mechanics*, 660:240–257, 2010.

J. Gula, V. Zeitlin, and R. Plougonven. Instabilities of two-layer shallow-water flows with vertical shear in the rotating annulus. *Journal of Fluid Mechanics*, 638:27–47, 2009.

G. Hadley. Vi. concerning the cause of the general trade-winds. *Philosophical Transactions of the Royal Society of London*, 39(437):58–62, 1735.

U. Harlander. Lecture notes in waves in fluids. unpublished, Januar 2014.

U. Harlander, T. von Larcher, Y. Wang, and C. Egbers. PIV-and LDV-measurements of baroclinic wave interactions in a thermally driven rotating annulus. *Experiments in Fluids*, 51(1): 37–49, 2011.

J. Hart. A laboratory study of baroclinic instability. *Geophysical & Astrophysical Fluid Dynamics*, 3(1):181–209, 1972.

D. Hathaway and W. Fowlis. Flow regimes in a shallow rotating cylindrical annulus with temperature gradients imposed on the horizontal boundaries. *Journal of Fluid Mechanics*, 172: 401–418, 1986.

I. M. Held and A. Y. Hou. Nonlinear axially symmetric circulations in a nearly inviscid atmosphere. *Journal of the Atmospheric Sciences*, 37(3):515–533, 1980.

A. Hertzog, C. Souprayen, and A. Hauchecorne. Observation and backward trajectory of an inertio-gravity wave in the lower stratosphere. In *Annales Geophysicae*, volume 19, pages 1141–1155, 2001.

R. Hide. An experimental study of thermal convection in a rotating liquid. *Philosophical Transactions of the Royal Society of London A: Mathematical, Physical and Engineering Sciences*, 250(983):441–478, 1958.

R. Hide. Theory of axisymmetric thermal convection in a rotating fluid annulus. *The Physics of Fluids*, 10(1):56–68, 1967.

R. Hide. Some laboratory experiments on free thermal convection in a rotating fluid subject to a horizontal temperature gradient and their relation to the theory of the global atmospheric circulation. *The global circulation of the atmosphere*, 1969.

R. Hide and W. Fowlis. Thermal convection in a rotating annulus of liquid: effect of viscosity on the transition between axisymmetric and non-axisymmetric flow regimes. *Journal of the Atmospheric Sciences*, 22(5):541–558, 1965.

R. Hide and P. Mason. Baroclinic Waves in a Rotating Fluid Subject to Internal Heating. *Philosophical Transactions of the Royal Society of London Series A*, 268:201–232, 1970.

R. Hide and P. Mason. Sloping convection in a rotating fluid. *Advances in Physics*, 24(1): 47–100, 1975.

R. Hide, P. Mason, and R. Plumb. Thermal convection in a rotating fluid subject to a horizontal temperature gradient: spatial and temporal characteristics of fully developed baroclinic waves. *Journal of the Atmospheric Sciences*, 34(6):930–950, 1977.

S. Hien, J. Rolland, S. Borchert, L. Schoon, C. Zülicke, and U. Achatz. Spontaneous inertia–gravity wave emission in the differentially heated rotating annulus experiment. *Journal of Fluid Mechanics*, 838:5–41, 2018.

P. Hignett. Characteristics of amplitude vacillation in a differentially heated rotating fluid annulus. *Geophysical & Astrophysical Fluid Dynamics*, 31(3-4):247–281, 1985.

P. Hignett, A. White, R. Carter, W. Jackson, and R. Small. A comparison of laboratory measurements and numerical simulations of baroclinic wave flows in a rotating cylindrical annulus. *Quarterly Journal of the Royal Meteorological Society*, 111(467):131–154, 1985.

C. O. Hines. Internal atmospheric gravity waves at ionospheric heights. *Canadian Journal of Physics*, 38(11):1441–1481, 1960.

J. R. Holton. An introduction to dynamic meteorology. *American Journal of Physics*, 41(5): 752–754, 1973.

T. Jacoby, P. Read, P. D. Williams, and R. Young. Generation of inertia–gravity waves in the rotating thermal annulus by a localised boundary layer instability. *Geophysical & Astrophysical Fluid Dynamics*, 105(2-3):161–181, 2011.

H. A. Kafiabad and P. Bartello. Rotating stratified turbulence and the slow manifold. *Computers & Fluids*, 151:23–34, 2017.

H. A. Kafiabad and P. Bartello. Spontaneous imbalance in the non-hydrostatic boussinesq equations. *Journal of Fluid Mechanics*, 847:614–643, 2018.

C. B. Ketchum. An experimental study of baroclinic annulus waves at large Taylor number. *Journal of the Atmospheric Sciences*, 29(4):665–679, 1972.

S. Khaykin, A. Hauchecorne, N. Mzé, and P. Keckhut. Seasonal variation of gravity wave activity at midlatitudes from 7 years of cosmic gps and rayleigh lidar temperature observations. *Geophysical Research Letters*, 42(4):1251–1258, 2015.

Y.-J. Kim, S. D. Eckermann, and H.-Y. Chun. An overview of the past, present and future of gravity-wave drag parametrization for numerical climate and weather prediction models. *Atmosphere-Ocean*, 41(1):65–98, 2003.

L. Lacaze, P. Le Gal, and S. Le Dizès. Elliptical instability in a rotating spheroid. *Journal of Fluid Mechanics*, 505:1–22, 2004.

R. Lagrange, P. Meunier, F. Nadal, and C. Eloy. Precessional instability of a fluid cylinder. *Journal of Fluid Mechanics*, 666:104–145, 2011.

R. Lambert and H. Snyder. Experiments on the effect of horizontal shear and change of aspect ratio on convective flow in a rotating annulus. *Journal of Geophysical Research*, 71(22):5225–5234, 1966.

M. Le Bars, D. Lecoanet, S. Perrard, A. Ribeiro, L. Rodet, J. M. Aurnou, and P. Le Gal. Experimental study of internal wave generation by convection in water. *Fluid Dynamics Research*, 47(4):045502, 2015.

C. Leith. Nonlinear normal mode initialization and quasi-geostrophic theory. *Journal of the Atmospheric Sciences*, 37(5):958–968, 1980.

M. J. Lighthill. On sound generated aerodynamically I. General theory. *Proc. R. Soc. Lond. A*, 211(1107):564–587, 1952.

D. K. Lilly. Stratified turbulence and the mesoscale variability of the atmosphere. *Journal of the Atmospheric Sciences*, 40(3):749–761, 1983.

E. Lindborg. The effect of rotation on the mesoscale energy cascade in the free atmosphere. *Geophysical research letters*, 32(1), 2005.

E. Lindborg. Horizontal wavenumber spectra of vertical vorticity and horizontal divergence in the upper troposphere and lower stratosphere. *Journal of the atmospheric sciences*, 64(3):1017–1025, 2007.

R. S. Lindzen and J. R. Holton. A theory of the quasi-biennial oscillation. *Journal of the Atmospheric Sciences*, 25(6):1095–1107, 1968.

E. N. Lorenz. Empirical orthogonal functions and statistical weather prediction. *Scientific Report No. 1. Department of Meteorology, MIT*, 1956.

E. N. Lorenz. Attractor sets and quasi-geostrophic equilibrium. *Journal of the Atmospheric Sciences*, 37(8):1685–1699, 1980.

A. Lovegrove, P. Read, and C. Richards. Generation of inertia-gravity waves in a baroclinically unstable fluid. *Quarterly Journal of the Royal Meteorological Society*, 126(570):3233–3254, 2000.

P. Lynch. *The emergence of numerical weather prediction: Richardson's dream*. Cambridge University Press, 2006.

P. Markowski and Y. Richardson. *Mesoscale meteorology in midlatitudes*, volume 2. John Wiley & Sons, 2011.

L. Marple. Computing the discrete-time "analytic" signal via FFT. *IEEE Transactions on signal processing*, 47(9):2600–2603, 1999.

J. W. Miles and H. E. Huppert. Lee waves in a stratified flow. Part 2. Semi-circular obstacle. *Journal of Fluid Mechanics*, 33(4):803–814, 1968.

T. L. Miller and W. W. Fowlis. Laboratory experiments in a baroclinic annulus with heating and cooling on the horizontal boundaries. *Geophysical & Astrophysical Fluid Dynamics*, 34 (1-4):283–300, 1985.

J. Mougel, D. Fabre, and L. Lacaze. Waves in Newton's bucket. *Journal of Fluid Mechanics*, 783:211–250, 2015.

W. Munk. Internal waves and small-scale processes. *Evolution of physical oceanography*, pages 264–291, 1981.

C. J. Nappo. *An introduction to atmospheric gravity waves.* Academic press, 2013.

NASA Earth Observatory. Wave clouds near amsterdam island, 2005. URL `https://earthobservatory.nasa.gov/images/6151/wave-clouds-near-amsterdam-island`. [Online; accessed June 22, 2019].

NASA Earth Observatory. Gravity waves generated by rapidly rising deep convection over the ocean, 2009. URL `https://www.nasa.gov/images/content/365848main_waveclouds-516.jpg`. [Online; accessed June 22, 2019].

NASA/GSFC. Animation of the jet stream from nasa's goddard space flight center, 2012. URL `https://www.nasa.gov/images/content/629341main_Earth_jet_stream.jpg`. [Online; accessed June 22, 2019].

G. Nastrom and K. S. Gage. A climatology of atmospheric wavenumber spectra of wind and temperature observed by commercial aircraft. *Journal of the atmospheric sciences*, 42(9):950–960, 1985.

G. Nastrom, K. Gage, and W. Jasperson. Kinetic energy spectrum of large-and mesoscale atmospheric processes. *Nature*, 310(5972):36, 1984.

A. Navarra and V. Simoncini. *A guide to empirical orthogonal functions for climate data analysis.* Springer Science & Business Media, 2010.

E. J. O'Neil. The stability of flows in a differentially heated rotating fluid system with rigid bottom and free top. *Studies in Applied Mathematics*, 48(3):227–256, 1969.

G. Oster and M. Yamamoto. Density Gradient Techniques. *Chemical Reviews*, 63(3):257–268, 1963.

D. O'Sullivan and T. J. Dunkerton. Generation of inertia–gravity waves in a simulated life cycle of baroclinic instability. *Journal of the Atmospheric Sciences*, 52(21):3695–3716, 1995.

J. Pedlosky. Finite-amplitude baroclinic waves. *Journal of the Atmospheric Sciences*, 27(1):15–30, 1970.

J. Pedlosky. Geophysical fluid dynamics. *New York and Berlin, Springer-Verlag, 1982. 636 p.*, 1982.

J. Pedlosky. *Waves in the ocean and atmosphere: introduction to wave dynamics.* Springer Science & Business Media, 2013.

R. L. Pfeffer, J. Ahlquist, R. Kung, Y. Chang, and G. Li. A study of baroclinic wave behaviour over bottom topography using complex principal component analysis of experimental data. *Journal of the Atmospheric Sciences*, 47(1):67–81, 1990.

N. A. Phillips. Energy transformations and meridional circulations associated with simple baroclinic waves in a two-level, quasi-geostrophic model. *Tellus*, 6(3):274–286, 1954.

R. Plougonven and C. Snyder. Gravity waves excited by jets: Propagation versus generation. *Geophysical research letters*, 32(18), 2005.

R. Plougonven and F. Zhang. Gravity waves: Gravity waves excited by jets and fronts. In *Encyclopedia of Atmospheric Sciences: Second Edition*, pages 164–170. Elsevier Inc., 2014a.

R. Plougonven and F. Zhang. Internal gravity waves from atmospheric jets and fronts. *Reviews of Geophysics*, 52(1):33–76, 2014b.

R. A. Plumb. Stratospheric transport. *Journal of the Meteorological Society of Japan. Ser. II*, 80(4B):793–809, 2002.

W. H. Press, S. A. Teukolsky, W. T. Vetterling, and B. P. Flannery. *Numerical recipes 3rd edition: The art of scientific computing*. Cambridge university press, 2007.

M. Raffel, C. E. Willert, F. Scarano, C. J. Kähler, S. T. Wereley, and J. Kompenhans. *Particle image velocimetry: a practical guide*. Springer, 2018.

A. Randriamampianina and E. C. del Arco. Inertia–gravity waves in a liquid-filled, differentially heated, rotating annulus. *Journal of Fluid Mechanics*, 782:144–177, 2015.

P. Read. Applications of singular systems analysis to 'baroclinic chaos'. *Physica D: Nonlinear Phenomena*, 58(1-4):455–468, 1992.

P. L. Read, E. P. Pérez, I. M. Moroz, R. M. Young, T. von Larcher, and P. Williams. General circulation of planetary atmospheres: Insights from rotating annulus and related experiments. *Modeling Atmospheric and Oceanic Flows: Insights from Laboratory Experiments and Numerical Simulations*, 205, 2014.

P. L. Read, F. Tabataba-Vakili, Y. Wang, P. Augier, E. Lindborg, A. Valeanu, and R. M. Young. Comparative terrestrial atmospheric circulation regimes in simplified global circulation models. part ii: Energy budgets and spectral transfers. *Quarterly Journal of the Royal Meteorological Society*, 144(717):2558–2576, 2018.

C. Rodda and U. Harlander. Properties of inertia-gravity waves emitted from jets and fronts in a atmosphere-like differentially heated rotating annulus. in preparation, 2019.

C. Rodda, I. Borcia, P. Le Gal, M. Vincze, and U. Harlander. Baroclinic, Kelvin and inertia-gravity waves in the barostrat instability experiment. *Geophysical & Astrophysical Fluid Dynamics*, pages 1–32, 2018.

C. Rodda, S. Hien, U. Achatz, and U. Harlander. A new atmospheric-like differentially heated rotating annulus configuration to study gravity wave emission from jets and fronts. submitted to Experiments in Fluids, 2019.

K. Sato. A statistical study of the structure, saturation and sources of inertio-gravity waves in the lower stratosphere observed with the mu radar. *Journal of atmospheric and terrestrial physics*, 56(6):755–774, 1994.

W. J. Saucier. Horizontal deformation in atmospheric motion. *Eos, Transactions American Geophysical Union*, 34(5):709–719, 1953.

F. Scarano and M. L. Riethmuller. Advances in iterative multigrid PIV image processing. *Experiments in Fluids*, 29(1):S051–S060, 2000.

H. Scolan and P. L. Read. A rotating annulus driven by localized convective forcing: a new atmosphere-like experiment. *Experiments in Fluids*, 58(6):75, 2017.

R. S. Scorer. Theory of waves in the lee of mountains. *Quarterly Journal of the Royal Meteorological Society*, 75(323):41–56, 1949.

R. Sharman, S. Trier, T. Lane, and J. Doyle. Sources and dynamics of turbulence in the upper troposphere and lower stratosphere: A review. *Geophysical Research Letters*, 39(12), 2012.

W. D. Smyth and J. R. Carpenter. *Instability in Geophysical Flows*. Cambridge University Press, 2019.

C. Snyder, D. J. Muraki, R. Plougonven, and F. Zhang. Inertia–gravity waves generated within a dipole vortex. *Journal of the Atmospheric Sciences*, 64(12):4417–4431, 2007.

J. Sommeria. Correlation imaging velocimetry at the coriolis facility. http://servforge.legi.grenoble-inp.fr/projects/soft-uvmat/attachment/wiki/WikiStart/CIV_doc_lim.pdf, February 2003.

N. Sugimoto, K. Ishioka, and K. Ishii. Parameter sweep experiments on spontaneous gravity wave radiation from unsteady rotational flow in an f-plane shallow water system. *Journal of the Atmospheric Sciences*, 65(1):235–249, 2008.

B. R. Sutherland. *Internal gravity waves*. Cambridge University Press, 2010.

B. R. Sutherland, U. Achatz, P. C. Colm-cille, and J. M. Klymak. Recent progress in modeling imbalance in the atmosphere and ocean. *Physical Review Fluids*, 4(1):010501, 2019.

S. Suzuki, K. Shiokawa, Y. Otsuka, S. Kawamura, and Y. Murayama. Evidence of gravity wave ducting in the mesopause region from airglow network observations. *Geophysical Research Letters*, 40(3):601–605, 2013.

L. Thomas, R. Worthington, and A. McDonald. Inertia-gravity waves in the troposphere and lower stratosphere associated with a jet stream exit region. In *Annales Geophysicae*, volume 17, pages 115–121. Springer, 1998.

J. Thomson. Bakerian lecture.—On the grand currents of atmospheric circulation. *Proceedings of the Royal Society of London*, 51(308-314):42–46, 1892.

R. E. Thomson and W. J. Emery. *Data Analysis Methods in Physical Oceanography*. Elsevier, 2001.

S. Triana, D. Zimmerman, and D. Lathrop. Precessional states in a laboratory model of the Earth's core. *Journal of Geophysical Research: Solid Earth*, 117.B4(B4), 2012.

L. W. Uccellini and S. E. Koch. The synoptic setting and possible energy sources for mesoscale wave disturbances. *Monthly weather review*, 115(3):721–729, 1987.

G. K. Vallis. *Atmospheric and oceanic fluid dynamics: fundamentals and large-scale circulation*. Cambridge University Press, 2006.

J. Vanneste. Balance and spontaneous wave generation in geophysical flows. *Annual Review of Fluid Mechanics*, 45, 2013.

J. Vanneste and I. Yavneh. Exponentially small inertia–gravity waves and the breakdown of quasigeostrophic balance. *Journal of the atmospheric sciences*, 61(2):211–223, 2004.

J. Vanneste and I. Yavneh. Unbalanced instabilities of rapidly rotating stratified shear flows. *Journal of Fluid Mechanics*, 584:373–396, 2007.

T. VanZandt. A universal spectrum of buoyancy waves in the atmosphere. *Geophysical Research Letters*, 9(5):575–578, 1982.

F. Vettin. Ueber den aufsteigenden Luftstrom, die Entstehung des Hagels und der Wirbel-Stürme. *Annalen der Physik*, 178(10):246–255, 1857.

M. Vincze and I. M. Jánosi. Laboratory experiments on large-scale geophysical flows. In *The Fluid Dynamics of Climate*, pages 61–94. Springer, 2016.

M. Vincze, U. Harlander, T. von Larcher, and C. Egbers. An experimental study of regime transitions in a differentially heated baroclinic annulus with flat and sloping bottom topographies. *Nonlinear Processes in Geophysics*, 21:237–250, 2014.

M. Vincze, S. Borchert, U. Achatz, T. von Larcher, M. Baumann, C. Liersch, S. Remmler, T. Beck, K. D. Alexandrov, C. Egbers, et al. Benchmarking in a rotating annulus: a comparative experimental and numerical study of baroclinic wave dynamics. *Meteorologische Zeitschrift*, pages 611–635, 2015.

M. Vincze, I. Borcia, U. Harlander, and P. Le Gal. Double-diffusive convection and baroclinic instability in a differentially heated and initially stratified rotating system: the barostrat instability. *Fluid Dynamics Research*, 48(6):061414, 2016.

Á. Viúdez. The origin of the stationary frontal wave packet spontaneously generated in rotating stratified vortex dipoles. *Journal of Fluid Mechanics*, 593:359–383, 2007.

Á. Viúdez and D. G. Dritschel. Spontaneous generation of inertia–gravity wave packets by balanced geophysical flows. *Journal of Fluid Mechanics*, 553:107–117, 2006.

T. Von Larcher and C. Egbers. Experiments on transitions of baroclinic waves in a differentially heated rotating annulus. *Nonlinear Processes in Geophysics*, 12(6):1033–1041, 2005.

T. Von Larcher, S. Viazzo, U. Harlander, M. Vincze, and A. Randriamampianina. Instabilities and small-scale waves within the Stewartson layers of a thermally driven rotating annulus. *Journal of Fluid Mechanics*, 841:380–407, 2018.

J.-S. von Storch, G. Badin, and M. Oliver. The interior energy pathway: Inertia-gravity wave emission by oceanic flows. In *Energy Transfers in Atmosphere and Ocean*, pages 53–85. Springer, 2019.

M. L. Waite and C. Snyder. The mesoscale kinetic energy spectrum of a baroclinic life cycle. *Journal of the Atmospheric Sciences*, 66(4):883–901, 2009.

J. M. Wallace and P. V. Hobbs. *Atmospheric science: an introductory survey*, volume 92. Elsevier, 2006.

S. Wang, F. Zhang, and C. Snyder. Generation and propagation of inertia–gravity waves from vortex dipoles and jets. *Journal of the Atmospheric Sciences*, 66(5):1294–1314, 2009.

T. Warn, O. Bokhove, T. Shepherd, and G. Vallis. Rossby number expansions, slaving principles, and balance dynamics. *Quarterly Journal of the Royal Meteorological Society*, 121(523):723–739, 1995.

J. Wei and F. Zhang. Mesoscale gravity waves in moist baroclinic jet–front systems. *Journal of the Atmospheric Sciences*, 71(3):929–952, 2014.

J. Wei and F. Zhang. Tracking gravity waves in moist baroclinic jet-front systems. *Journal of Advances in Modeling Earth Systems*, 7(1):67–91, 2015.

P. D. Williams, T. W. Haine, and P. L. Read. On the generation mechanisms of short-scale unbalanced modes in rotating two-layer flows with vertical shear. *Journal of Fluid Mechanics*, 528:1–22, 2005.

P. D. Williams, T. W. Haine, and P. L. Read. Inertia–gravity waves emitted from balanced flow: Observations, properties, and consequences. *Journal of the atmospheric sciences*, 65(11): 3543–3556, 2008.

D. L. Wu and F. Zhang. A study of mesoscale gravity waves over the north atlantic with satellite observations and a mesoscale model. *Journal of Geophysical Research: Atmospheres*, 109(D22), 2004.

F. Zhang. Generation of mesoscale gravity waves in upper-tropospheric jet–front systems. *Journal of the atmospheric sciences*, 61(4):440–457, 2004.

Acknowledgement

During the time of my PhD, I met a great number of people that I would like to thank for their contribution to this thesis.

First and foremost, I am deeply grateful to my PhD supervisor Uwe Harlander for his patience and constant support. The kindness with which he shared his experience and vast knowledge about waves have made me really enjoy the time spent working together.

I also want to thank all members of the Spontaneous Imbalance research group, in particular Ulrich Achatz, Christoph Zülicke, Steffen Hien, Lena Schoon, and Ion Borcia for the work done together and the fruitful discussions throughout my PhD. I am greatly thankful to the DFG for founding my position within the MSGWaves project. A particular thanks to Aurelia Müller for the enthusiasm with which she organised several activities for the female young scientists and, in particular, for the coaching sessions with Julie Sterns. I'm also grateful to Chantal Staquet and Joel Sommeria for hosting me at LEGI in Grenoble and to Patrice Le Gal for inviting me to IRPHE in Marseille. My research profoundly benefited from both visits.

My time at the BTU Cottbus wouldn't have been the same without my colleagues in the aerodynamics and fluid mechanics department. Thank you all for making me feel welcome and for the great time together. A special thanks goes to Torsten Seelig, Antoine Meyer, Wenchao Xu, and Michael Hoff for finding the time to read part of my work and the useful comments about this manuscript. I would also like to acknowledge the work of the students I supervised in the laboratory: Dumitru Sandu, Lavinia –Nicoleta Aparaschivei, Tudor Cimpeanu, Tito Rodda, and Wenying Zu.

Finally, a huge thanks to my family for their love and support from very far away and not too far away. And I would also like to thank Stephen Budd not only for sharing with me good and bad times, but also for the time spent proofreading some of my work and patiently helping me improving my english!

www.ingramcontent.com/pod-product-compliance
Lightning Source LLC
Chambersburg PA
CBHW070717220326
41598CB00024BA/3203